Emergency
Responders
Guide to
Chemical
Reactivity and
Compatibility

EMERGENCY RESPONDERS GUIDE TO CHEMICAL REACTIVITY AND COMPATIBILITY

Donald Drum, B.S., M.S., Ph.D.

McGraw-Hill

New York   Chicago   San Francisco   Lisbon   London   Madrid
Mexico City   Milan   New Delhi   San Juan   Seoul
Singapore   Sydney   Toronto

Cataloging-in-Publication Data is on file with the Library of Congress

McGraw-Hill

A Division of The McGraw-Hill Companies

1 2 3 4 5 6 7 8 9 0 DOC/DOC 0 9 8 7 6 5 4 3 2

ISBN 0-07-138900-8

The sponsoring editor for this book was Kenneth P. McCombs, the editing supervisor was Steven Melvin, and the production supervisor was Sherri Souffrance. It was set in the CHB01 design in Times Roman by Joanne Morbit of the McGraw-Hill Professional's Hightstown, New Jersey composition unit.

Printed and bound by R. R. Donnelley & Sons Company.

McGraw-Hill books are available at special quantity discounts to use as premiums and sales promotions, or for use in corporate training programs. For more information, please write to the Director of Special Sales, Professional Publishing, McGraw-Hill, Two Penn Plaza, New York, NY 10121-2298. Or contact your local bookstore.

This book was printed on recycled, acid-free paper containing a minimum of 50% recycled, de-inked fiber.

To all of the emergency response teams, site workers, law enforcement officers, safety personnel, and individuals who have addressed the disasters at the Federal Building in Oklahoma City, the World Trade Center in New York City, the Pentagon in Washington, D.C., and the plane crash in Pennsylvania. The author hopes that this text will aid in the prevention and the solving of various types of accidents involving chemicals, storage problems, chemical handling issues, and crimes.

CONTENTS

CONTRIBUTORS

Deborah Bowers   Research and Development, Jacksonville, Florida.

James Chvala   Hazardous Materials Specialist, Local Emergency Planning Committee, Butler County, Butler, Pennsylvania.

Charles Craig   Former Hazardous Materials Specialist, Local Emergency Planning Committee, Butler County, Butler, Pennsylvania.

David Drum   Research and Development, Independence, Missouri.

Paul Harrity   II-VI, Inc., Saxonburg, Pennsylvania, and Deputy Commander of Butler County Hazmat Team 100, Butler, Pennsylvania.

Greg Haughey   Mine Safety Appliances, Evans City, Pennsylvania, and Local Emergency Planning Committee, Hazmat Instructor and Commander of Butler County Hazmat Team 100, Butler, Pennsylvania.

George Janson   Instructor (1978–1988), Hazardous Waste Seminars, Columbia-Greene Community College, Hudson, New York.

Harry Peterson   Research and Development, Butler, Pennsylvania.

Ben Pierson, P.E.   Senior Sanitary Engineer, State of New York Department of Health, Troy, New York.

Natalie J. Price   Laboratory Coordinator, Butler County Community College, Butler, Pennsylvania.

Larry Slagle   Coordinator of Public Safety and Hazardous Materials Training, Butler County Community College, Butler, Pennsylvania.

George Smith   Fire Training Coordinator, Local Emergency Planning Committee, Butler County, Butler, Pennsylvania.

Lynn L. Thompson   Professor of Natural Science and Technology, Butler County Community College, Butler, Pennsylvania.

National Fire Protection Association   1 Batterymarch Park, P.O. Box 9101, Quincy, Massachusetts 02269.

United States Consumer Products Safety Commission   4330 East West Highway, Bethesda, Maryland 20814.

PREFACE

When I approach the scene of an accident, how do I know if the combination of chemicals will be dangerous? If two or more chemicals accidentally mix or come into contact with each other, will the mixture catch fire, produce toxic gases, explode, or form dangerous material?

The shipping papers, container labels, placards, and emergency response guides alone are inadequate to predict the outcome of a *combination* of substances. This is the first manual designed specifically for predicting the results of mixing two or more substances and/or chemicals which can result in accidents and injury to individuals, animals, and property. The tables can be used to prevent potential mixing of hazardous combinations of materials in storage or transportation. Once an accident has occurred it assists the responders in making rapid and appropriate decisions to limit the impact of biological and chemical toxins.

This guide draws upon information and experiences shared during seminars over 15 years with people from many different industries, communities, and countries. The textbook has practical training and response applications for members of the international workforce including, but not limited to, emergency responders, anti-terrorist officials, investigators, inspection specialists, medical and professional staff, government and law enforcement officials, employees of chemical and waste treatment facilities, transportation industry, and company employees.

If the chemicals or their gases in the appropriate concentrations and states interact, this handbook will predict the consequences and events that will likely occur. Incompatible chemicals and wastes can produce a number of undesirable or uncontrolled conditions such as explosions, fire, flammable gases, heat, toxic chemicals (gas, vapors, liquid, or solid), shock- and friction-sensitive compounds, solubilization of toxic materials, pressurization of closed containers, violent polymerization, decomposition, and toxic dusts, particles, and mists. The events of a reaction can change dramatically depending upon changes in mixing or environmental conditions, changes in chemical concentrations, or presence of chemical additives. The consequences of mixing incompatible chemicals can include personal injury or death, property and equipment damage, air and water contamination, and vegetation and environmental damage. In addition, the presence of biosensitive materials can produce air and water contamination leading to serious illness or death.

The compounds and substances in the lists demonstrate a variety of reactivity levels in a fire, explosion, or disastrous event. The lists of compounds also include many of the more toxic forms of herbicides, insecticides, rodenticides, and chemicals (i.e., nematicides, fumigants, plant growth regulators, fungicides, acaricides, larvicides, wood preservatives, ovicides, avicides, and aphicides). Common compounds with limited toxicity level are also included in these lists because of the problems created during a fire or some other catastrophic event. Many of the chemicals are sold in more than one country under a variety of names, while some chemicals have greater application in specific countries with limited application in the United States.

Different trade, chemical, and acronym names for the same compound are listed with identical reactivity group numbers, or RGNs. For instance, PCB and polychlorinated biphenyl are the same compound with different names and are listed with the same RGNs. Likewise, hydrochloric acid and muriatic acid are different names for the same compound and have equivalent RGNs. Also, benzoepin, endosulfan, and Thiodan* are common names for the same compound and have identical RGNs.

Some gases with RGN 35 through 39 will not react directly with solids or liquids. However, some vapors from liquids or solids will form easily ignitable mixtures in the vapor phase. These mixtures and products of fire are usually toxic, flammable, and often explosive within specific ranges of concentration. Many of the codes (E, F, H, T, etc.) for RGN 35 to 39 are expressed in *italics* in the "Chemical Reactivity and Compatibility Chart" to represent gaseous reactions. Additional codes in normal print indicate reaction components in *any* state or phase. In addtion, two compounds in the presence of each other may not normally react to cause a fire, explosion, or other event. However, if the two compounds are present during a fire or an explosive event, the compounds in the volatized state may react in a different manner than expected. Not all possible reaction factors or conditions are represented or are present in the "Chemical Reactivity and Compatibility Chart."

In this text, explosive materials can be defined, in varying degrees, from those chemicals that give off a little "pop" and do almost no property damage to those chemicals that create an event that severely damages the surroundings. Many of the more dangerous materials can also be found in more specific detail in other literature sources. The reaction conditions or concentrations of components for an explosion are not defined in this text. This information is intended *only* as a guide and specific conditions may alter the actual intensity of the event.

Increases in the concentrations of chemicals or materials in an enclosed space, such as a building or sewer line, can produce an explosion or catastrophic event. The explosion will be dependent on concentrations of chemicals, oxygen availability, sparking or igniting events, moisture content, gas concentrations, pressure, temperature, sunlight, and other factors. *Not* every catastrophic event or explosion has been predicted in this document.

After a little practice, chemical characterization from accidents or spills of two or three components can be easily defined in a few minutes. Most accidents occur with two or three components. Accidents and spills or leaks with many components can occur at industrial sites, landfills, waste and chemical storage, recovery centers, and during transportation.

This guide has organized chemicals and substances into groups which have similar reaction characteristics and properties. These general groupings of reactions assist us in preventing accidents and responding quickly to an event. Because of the infinite possible combinations of chemicals and conditions, the text cannot be considered to be complete and accurate under all circumstances and should not be considered to be a substitute for thorough training in chemical and biological emergency response. The author has made every attempt to present all possible broad categories of conditions and chemicals accurately. The author requests that he be contacted concerning any additions or corrections in the text: Donald A. Drum, Ph.D., 630 Brice Road, Reynoldsburg, Ohio 43068. Each potential change or addition will be evaluated seriously before updates are made.

ACKNOWLEDGMENTS

The author wishes to express his grateful appreciation to the contributors for their assistance in supplying information and materials for this text. He expresses his appreciation to his wife, Pamela Drum, for her support and efforts during the development of the text. Without the support and encouragement of the contributors and family members, this task would not have been completed.

The author wishes to express his appreciation to Scott Grillo, Editor-in-Chief, and Ken McCombs, Acquisition Editor, Professional and Reference Division, McGraw-Hill, New York, for their encouragement and invaluable guidance during the successful completion of this text. The staff at McGraw-Hill, Inc., has been extremely helpful during the printing of the material.

PROCEDURE TO IDENTIFY CHEMICAL INCOMPATIBILITY

The stepwise procedure is provided to avoid accidents and potential problems when two or more different chemicals or waste streams are mixed. **The user of these materials should be aware that *not* all potential accidents or problems can be avoided by the application of this procedure.** To identify potential problems among chemicals and/or biological materials being used in a given environment:

1. Identify the chemicals or materials in the potential spill, accident, or mix through the hauling papers, manifest, testing, formulation documents, or other information sources.

2. Separately list the individual chemicals (materials) of each mixture of component(s) involved in the spill, accident, or mix of chemicals.

Note: Components of one mixture are listed individually and separately from those of another mixture.

3. Find each chemical in the "Alphabetical List of Compounds and RGNs" (first table in the text). For each chemical list the corresponding RGN (number or numbers) separately. The RGNs for chemicals in one mixture are listed separately from the RGNs of chemicals for the second mixture.

4. Using the "Chemical Reactivity and Compatibility Chart," locate in the vertical column (on the left-hand side of the table) the RGN for the first chemical in the potential spill. Then, locate in the horizontal row (across the bottom of the table) the RGN for the second chemical listed separately which has the potential to come in contact with the first chemical. Using the chart, find the intersection of the row and the column. Continue this procedure for each chemical in the two different mixtures (listed individually and separately).

5. Read and record the symbols or codes from the intersection of the row and column in the "Chemical Reactivity and Compatibility Chart." A number of symbols, such as F, G, T, E, etc., will be recorded for the two chemicals. Repeat the process of recording the codes or symbols for any other chemicals that may be mixed due to a spill or accident.

6. Identify the characteristics of mixing two or more incompatible chemicals by locating the codes or symbols (F, G, T, E, etc.) in a short definition listing provided on the chart or in the table titled "Chemical Incompatibility Codes."

7. In addition, the reaction characteristics of each chemical can be identified in the back of this text within the section titled "Chemical Class and Chemical Reactivity Characteristics." After identifying the RGN(s) for a specific chemical from the first table in the text, proceed to the group number in the back of the text and refer to the description. For example, Caro's acid in the "Alphabetical List of Compounds and RGNs" is listed in RGNs of 2, 64. Proceed to groups RGN 2 and RGN 64 in the back of the text and read the descriptions preceding the list of similar compounds.

CHEMICAL INCOMPATIBILITY CODES

Description	CODE
Exposion, resulting from vigorous reactions producing sufficient heat (or shock) to detonate unstable components	E
Flammable/fire, due to ignition or potential ignition of reaction mixture; substances ignite easily and burn easily	F
Gases (carbon dioxide, nitrogen, generally nontoxic gases), produced by reactions	G
Heat, generated by chemical reactions	H

Toxic gases are gases which are extremely toxic to humans and animals. Such gases include fluorine, chlorine, phosgene, hydrogen cyanide, sulfur oxides, hydrogen sulfide, and nitrogen oxides.   **T**

Violently decompose are substances that tend to be heat and shock sensitive   **V**

Polymerization produces a macromolecule which is composed of a number of monomers; the linking together of smaller units (monomers) to produce long-chain polymers   **P**

Solubilization of toxic or very dangerous substances   **S**

Unknown characterization of reactions and substances   **U**

NOTES:

1. Symbols in *parenthesis* indicate that some of the chemicals in a particular group will interact with some of the chemicals in the second group.

2. Symbols in *italics* in the chart indicate chemical gaseous phase.

APPLICATIONS AND EXAMPLES

1. A laboratory technician accidentally pours a bottle of alcohol (ethanol) into a container of sulfuric acid. What may happen if the two chemicals are sufficiently concentrated to cause problems?

   SOLUTION: First, identify the RGNs for ethanol and sulfuric acid. See the "Alphabetical List of Compounds and RGNs" located in the front of the text.

   Solution A: RGN (sulfuric acid) = 2, 64, 67
   Solution B: RGN (ethanol) = 4

   Then, find the "Chemical Reactivity and Compatibility Chart" and locate the intersection of column 4 and row 2. The results are H, G, (E). The intersection of column 4 and row 64 yields H, (E), F, (T). Also, the results of the intersection of column 4 and row 67 are E, (T), F. Again, the definitions for the symbols are provided on the same page under the chart or on the page titled "Chemical Incompatibility Codes." The final results of the potential reaction are H, G, E, (T), F. Therefore, the predicted results are H = heat generated, G = gas given off, E = explosion potential, T = toxic substance may be produced, and F = fire potential.

4 PROCEDURE TO IDENTIFY CHEMICAL INCOMPATIBILITY

2. A person is attempting to clean a spot from the bathroom floor. He pours some ammonia cleaner identified as ammonium hydroxide on the floor, but the spot is not removed by rubbing. Then he pours a small amount of Chlorox solution (sodium hypochlorite) on the same spot. What will happen?

SOLUTION: The bottles or containers provide the list of active ingredients on the label. Then, identify the RGNs for ammonia cleaner (as ammonium hydroxide) and sodium hypochlorite using the "Alphabetical List of Compounds and RGNs" located in the front section of the text.

Solution A: RGN (Ammonium hydroxide) = 10, 39
Solution B: RGN (Sodium Hypochlorite) = 64

The intersection of column 10 and row 64 in the "Chemical Reactivity and Compatibility Chart" displays the letter T, which is defined as *toxic gas* by the code definitions under the "Chemical Incompatibility Codes." Also, column 39 and row 64 (or column 64 and row 39) indicate (T). Likewise, the intersection of row 10 and column 64 in the chart should provide the same results of T, or toxic gas. *Note: The results would have been the same if sodium hypochlorite in basic solution (RGN 10, 64) were compared to ammonium hydroxide (RGN 10, 39).*

3. Approximately 2 cups of a liquid drain cleaner containing sodium hypochlorite, sodium hydroxide, sodium silicate, and biodegradable surfactants are dumped down a clogged drain. The drain still remains plugged, so the homeowner pours dilute hydrochloric acid in water purchased from the local hardware store down the same drain. What results are predicted?

SOLUTION:

	Compound	RGN
Solution A	Sodium hypochlorite	64
	Sodium hydroxide	10
	Sodium silicate	Not listed (not really a problem)
	Biodegradable surfactants	Not listed
Solution B	Hydrochloric acid (in water) (not concentrated acid)	2

After looking at the different solutions and the RGNs (numbers), we would match the following: 2 versus 10 and 2 versus 64. The intersection of column 2 and row 10 in the "Chemical Reactivity and Compatibility Chart" indicates (F), E and the intersection of column 2 and row 64 suggests H, (V), (T).

After consulting the listing of code definitions ("Chemical Incompatibility Codes") of H, (V), (T), (F), and E, a number of serious events could occur. The immediate results of mixing dilute solutions would be H = heat and E = explosion (at least a sudden expansion of gases). The events would depend on the concentrations of the components in the final mixture. The most serious consequences of the codes predict: H = heat given off, V = violent decomposition, T = toxic gas emitted, F = fire possible, and E = explosive conditions. *Note: V, T, and F are in parenthesis which indicates that these activities may happen depending on concentrations. If a solution of concentrated HCl (with limited water) was mixed with solution A in a drain, then the results would be considerably more dangerous.*

4. A concentrated hydrochloric acid (muriatic acid) solution is accidentally spilled on a spot on a floor surface containing a strong cleaning solution. The industrial cleaning solution on the floor contains sodium hydroxide and sodium carbonate.

SOLUTION:

	Compound	RGN
Solution A	Hydrochloric acid (muriatic acid, concentrated)	2, 67
Solution B	Sodium hydroxide	10
	Sodium carbonate	10

After looking at the different solutions and RGNs, we would match the following: 2 versus 10 and 10 versus 67. The intersection of column 2 and row 10 in the "Chemical Reactivity and Compatibility Chart" indicates (F), E. Intersection of row 10 and column 67 indicates E, H, (F). Therefore, the mixing of the chemicals from the accidental spill would result in E, H, (F), where E = explosion, H = heat given off, and (F) = a potential fire.

5. Solution A containing butylamine (aminobutane) is accidentally contaminated with a solution of butyl peroxide. What results are expected?

SOLUTION: Using the "Alphabetical List of Compounds and RGNs" in the front of the text , we would find the RGNs for the specific compounds.

Solution A:	Butylamine	RGN = 7, 10
Solution B:	Butyl peroxide	RGN = 30

Then, we would match 7 versus 30 and 10 versus 30 in the "Chemical Reactivity and Compatibility Chart." Intersection of row 7 and column 30 (or row 30 and column 7) indicates H, T, P. Row 10 and column 30 (or row 30 and column 10) are blank in the chart.

Again, consult the chart titled "Chemical Incompatibility Codes" to identify the codes of H, T, P. Therefore, the predicted result of mixing the butylamine and butyl peroxide solutions would be H = heat given off, T = toxic gas emitted, and P = polymerization occurs.

NOTE: The chart titled "Chemical Reactivity and Compatibility Chart" is a predictive tool used to indicate what *may occur* when mixing two or more substances or chemicals. The predicted results *are not* the final conclusion for every possible situation. Each combination of chemicals will be influenced by a number of factors including, but not limited to, temperature, pressure, state of reactants, concentration of components, additional reactants or impurities, air and water (moisture) availability, unusual reaction conditions, and errors in the chart.

REFERENCES

Abram S. Benenson (ed.), *Control of Communicable Diseases in Man*, 15 th ed., American Public Health Association, Washington, DC, 1990.

Donald A. Drum, *Incompatibility of Hazardous Materials*, Columbia-Greene Community College, Hudson, NY, 1980, revised 1984 and 1988.

Donald Drum, et al., *Environmental Field Testing and Analysis Ready Reference Handbook*, McGraw-Hill, New York, 2001.

Gessner G. Hawley (ed.), *The Condensed Chemical Dictionary*, 10th ed., Van Nostrand Reinhold Company, Inc., New York, 1981.

Richard J. Lewis, Sr. (ed.), *Hawley's Condensed Chemical Dictionary*, 14th ed., John Wiley & Sons, Inc. (Wiley-Interscience), New York, 2001.

Richard T. Meister (editor in chief), *Farm Chemicals Handbook '98*, MeisterPRO Reference Guides, vol. 84, Meister Publishing Company, Willoughby, OH, 1998.

Material Safety Data Sheet, Genium Publishing Corp., Schenectady, NY, 1987.

The Report on Carcinogens, 9th ed., Section 301 (b) (4) of the Public Health Services Act, 2000.

2000 Emergency Response Guidebook, U.S. Department of Transportation, Washington, 2000.

The Material Safety Data Sheet (MSDS) and chemical fact sheets will provide some limited evidence of physical characteristics and chemical reaction. MSDS include some important information about incompatibility of a chemical with specific substances. These documents relate directly to use, handling, mixing, and proper disposal of pure chemicals, mixtures of chemicals, and chemical wastes.

EMERGENCY RESPONSE GUIDE TO CHEMICAL REACTIVITY AND COMPATIBILITY

ALPHABETICAL LIST OF COMPOUNDS AND RGNS

A

Name(s)	RGN
Acetylene tetrabromide	17
Acetyl iodide	17, 67
Acetyl ketene (Diketene)	63, 67
Acetylmethylcarbinol	19
Acetyl nitrate	27, 62
Acetyl peroxide	30, 62
Acetylsalicylic acid (Aspirin)	3, 13
Acid butyl phosphate	3
Acids, inorganic, n.o.s.	2
Acids, organic, n.o.s.	3
Acifat*	32
Acifon*	32
CAN (ACNQ)	7, 17, 19
Acridine	7
Acrolein	5, 63
Acrylamide (*RAHC*)	6, 40, 63
Acrylates, n.o.s.	13
Acrylic acid (with or without sulfuric acid)	3, 63
Acrylonitrile (Human carcinogen; *RAHC*)	26, 40, 63
Actinium (Radioactive poison)	23
Adamsite (Diphenylamine chloroarsine)	7, 24
Adipic acid	3
Adiponitrile	26
Aerozine (UDMH)	7, 62
Aflatoxins, B_2 and G_1 (Carcinogens)	40
Afugan*	32
Agent orange (Dioxin traces)	3, 17, 40
Aimcosystox*	20, 32
Alachlor	9, 17
Alanycarb	9, 20
Alcohol, n.o.s. (Ethanol, Isopropanol, Methanol, etc.)	4
Aldehyde ammonia	5, 7
Aldehyde, n.o.s.	5

Name(s)	RGN
para-Aminodiethylaniline	7
para-Aminodiethylaniline hydrochloride	7
para-Aminodimethylaniline	7
2-Amino-4,6-dinitrophenol	7, 27, 62
4-Aminodiphenyl (Human carcinogen)	7, 40
para-Aminodiphenylamine	7
Aminoethane (Ethylamine)	7, 10
Aminoethanol	4, 7
Aminoethoxyethanol (Diglycolamine*, DGA)	4, 7
Aminoethylpiperazine (*N*-Aminoethylpiperazine)	7
Aminoethylsulfuric acid	3
Amino-G acid	3
Amino-J acid	3
Aminomethane (Methylamine)	7, 10
1-Amino-2-methylanthraquinone (*RAHC*)	40
Aminomethylpropanol (AMP)	4, 7
Aminomethylpyridine (Various forms)	7
Aminomethylpyridine complexes	7
2-Aminonaphthalene (2-Naphthylamine; Human carcinogen)	7, 40
Aminopentane	7
Aminophenetole	7
Aminophenol (*o*-, *m*-, or *p*-)	7, 31
4-Amino-1-phenol-2,6-disulfonic acid	3
Aminophos	32
para-Aminophenylmercaptoacetic acid	3, 7
2-Amino-3-picoline (Aminopicoline complexes)	7
Aminopropane (Propylamine)	7, 10
Aminopropanol (2-Amino-1-propanol, 3-Amino-1-propanol)	4,7
Aminopropionic acid (Various forms)	3,7
Aminopropionitrile	7, 26
Aminopropyldiethanolamine	4, 7
Aminopropylmorpholine	7

Name(s)	RGN
Anisidine(s)...	7
o-Anisidine hydrochloride (*RAHC*)......................	7, 40
Anisole (Methylphenyl ether)............................	14
Anisoyl chloride...................................	17, 62, 67
Anisylacetone..	19
Anthio*...	32
Anthracene (Carcinogen)................................	16
Anthracene oil (Possible Carcinogen)...................	16
Anthranilic acid (*ortho*-Aminobenzoic acid)............	3, 7
Anthraquinone...	61
Anthrax (Various forms, Acute bacterial disease)......	40
Antifreeze..	4
Antimony..	22
Antimony-124 (Radioactive poison; Element).............	23
Antimony-125 (Radioactive poison; Isotope).............	23
Antimony chloride...................................	24, 67
Antimony compound, n.o.s..............................	24
Antimony-124 (Radioactive poison; Compounds)...........	23
Antimony fluoride................................	15, 24, 67
Antimony hydride (Stibine)...........................	24, 36
Antimony lactate......................................	24
Antimony nitride....................................	24, 25
Antimony oxide..	24
Antimony oxychloride..................................	24
Antimony pentachloride..............................	24, 67
Antimony pentafluoride............................	15, 24, 67
Antimony pentasulfide.............................	24, 33, 65
Antimony pentoxide....................................	24
Antimony perchlorate...............................	24, 64
Antimony postassium tartrate..........................	24
Antimony powder (Fine)................................	22
Antimony salt (deHaens salt)..........................	24
Antimony sulfate....................................	24, 67

Name(s)	RGN
Antimony sulfide (Antimony sulphide)	24, 33, 65
Antimony tribromide	24, 67
Antimony trichloride	24, 67
Antimony trifluoride	15, 24
Antimony triiodide	24, 67
Antimony trioxide (Human carcinogen)	24, 40
Antimony trisulfate (Antimony sulfate)	24
Antimony trisulfide (Antimony trisulphide)	24, 33, 65
Antimony trivinyl	24, 67
Antimony yellow (Naples yellow)	24
Antok*	4
ANTU	6
Apache*	32
Aqualin	5, 63
Aqua regia	2, 64, 67
Aramite* (Carcinogen)	17
Areginal*	13
Argon	1
Aroclor*or Arochlor* (Aroclor® 1254, Aroclor® 1260; *RAHC*)	17, 40
Arsanilic acid (*para*-Aminophenylarsonic acid)	3, 24
Arsenic (Human carcinogen and mutagen)	22, 40
Arsenic acid (Orthoarsenic acid, Arsenic pentoxide, CCA)	3, 24
Arsenical dust	22
Arsenical pesticide	24
Arsenic bromide (Arsenic tribromide)	24, 67
Arsenic chloride (Arsenic trichloride)	24, 67
Arsenic compound(s), inorganic, n.o.s. (Human carcinogen)	24, 40
Arsenic disulfide	24, 33, 65
Arsenic fluoride (Arsenic trifluoride)	15, 24, 66, 67
Arsenic iodide (Arsenic triiodide)	24, 67
Arsenic oxide	24

Name(s)	RGN

Arsenic pentafluoride, gas 15, 24, 39, 67
Arsenic pentaselenide 24
Arsenic pentasulfide 24, 33, 65
Arsenic pentoxide 24
Arsenic sulfide (Arsenic sulphide) 24, 33, 65
Arsenic tribromide 24, 67
Arsenic trichloride 24, 67
Arsenic trifluoride 15, 24, 66, 67
Arsenic triiodide 24, 67
Arsenic trioxide (Arsenious oxide; Human
   carcinogen) 24, 40
Arsenic trisulfide (Arsenic trisulphide) 24, 33, 65
Arsenious oxide 24
Arsine 24, 36, 65
Artic* 39
Aryl sulfonic acids (with and without sulfuric acid) 3
Asbestos dust particles (Human carcinogen) 40
Ascaridole 64
Askarel (Chlorinated organics) 17
Asphalt (petroleum) fumes 61
Aspirin dust 3, 62
Assert* 6, 13
Atrazine 7, 17
Atropine (Atropine sulfate) 7
Avadex* 12, 17
Avenge* 7
Avitrol* (4-Aminopyridine) 7
Axall* 17, 26
Azacitidine (5-Azacytidine) 40
Azelaic acid 3
Azide, n.o.s. 62
Azidithion (Menazon) 17, 32
Azidocarbonyl guanidine 8, 62

B

Name(s)	RGN
Bromophos	17, 32
Bromophosgene	17, 67
Bromopicrin	17, 27, 62
2-Bromopropane (All bromopropanes)	17
Bromopropene (All bromopropenes)	17
Bromopropionic acid	3, 17
Bromopropyne (All bromopropynes)	17
Bromosilane	17, 65
Bromosuccinic acid	3, 17
N-Bromosuccinimide, dry (NBS)	17, 66
Bromothiophenol	17, 20
Bromotoluene (Various forms)	17
Bromotrichloromethane	17
Bromotrifluoroether	17, 35
Bromotrifluoroethylene,gas (monomer) or liquid (polymer)	17, 35
Bromotrifluoromethane (Bromotrifluomethane)	17, 39
Bromoxynil	17, 26
Bronopol	4, 17, 27
Bronze powder	22
Broprodifacoum	4, 17, 19
Brotal*	17, 27
Brucine	6, 7
B.S. 500*	3
Bubonic plague (Bacterial infection)	40
Bufencarb	9
Bufotenine	4, 7
Bulan	17, 27
Bulldock*	17, 26
Busulfan (1,4-Butanediol dimethylsulfonate; Human carcinogen)	4, 40
Butachlor	6, 17
1,3-Butadiene, gas or liquid (Butadiene, Divinyl; Human carcinogen; May form explosive peroxides in air)	28, 35, 40, 62, 63, 66

C

Name(s)	RGN
Calcium nitrate	62, 64
Calcium nitride	25, 67
Calcium oxychloride	64
Calcium perchlorate	64
Calcium perchromate	64
Calcium permanganate	64
Calcium peroxide	64
Calcium phosphide	67
Calcium plumbate	64
Calcium propanearsonate (Calcium propyl arsonate)	24
Calcium resinate	24, 66
Calcium selenate	24
Calcium silicide (Decomposes in hot water)	66, 67
Calcium silicofluoride	15, 24, 67
Calcium silicon (Calcium-silicon alloy)	66
Calcium stannate	24
Calcium stearate dust	24
Calcium sulfate dust	24
Calcium sulfide	33, 65, 67
Calixin*	61
Calo-Clor*	24
Calocure*	24
Calogran*	24
Calomel (Mercurous chloride)	24
2-Camphanol	61
Camphechlor (*RAHC*)	17, 32, 40
Camphene	61
Camphor	19, 61
Camphor oil	61
Candit*	13
Cannabis	4
Capric acid	3
Caproic acid	3

Name(s)	RGN
Cellulose acetate	13
Cellulose compounds, fibers, pellets, granules, n.o.s.	13
Cellulose nitrate	27, 62
Ceramix*	24
Ceresan*	24
Ceric ammonium nitrate	64
Ceric sulfate	64
Cerium (Turnings or gritty powder)	21, 65
Cerium-141 (Radioactive poison, Compound)	23
Cerium hydride	65
Cerium-iron alloys (Potential pyrophoric)	21, 65
Cerium nitrate	64
Cerium trisulfide	33, 65
Cerous fluoride	15, 24
Cerous nitrate	64
Cerous oxalate	24
Cerous phosphide	65
Cesium	21
Cesium-137 (Radioactive poison)	23
Cesium amide	67
Cesium antimonide	24
Cesium arsenide	24
Cesium azide	62
Cesium carbide	65
Cesium fluoride	15
Cesium hexahydroaluminate	65
Cesium hydride	65, 67
Cesium hydroxide	10
Cesium nitrate	64
Cesium nitrite	64
Cesium perchlorate	64
Cesium peroxide	64
Cesium phosphide	67

Name(s)	RGN
Cesium sulfide	33, 65
Cesium tetroxide	64, 67
Cesium trioxide	64, 67
Cetyl mercaptan	20
Cetyl vinyl ether	14
Charcoal, wood	66
4-ChFu	3, 17
Chicago acid (SS acid)	3
Chinalphos	32
Chinmix*	17, 26
Chinomethionat	20
Chloracetyl chloride (Lachrymator)	17, 67
Chloral (Trichloroacetaldehyde)	5, 17
Chloral hydrate (Knockout drops)	4, 17
Chloralose (*alpha*-Chloralose)	17
Chloramben	3, 17
Chlorambucil (Human carcinogen)	3, 17, 40
Chloramine (Not to be confused with Chloramine-T)	17, 67
Chloramine-T	17, 64
Chloramizo	17
Chloramphenicol	4, 6, 17
Chloranil	17
Chlorate, inorganic, n.o.s	64
Chlorazine	7, 17
Chlorbenside	17, 20
Chlordane (Animal carcinogen)	17
Chlordecone (*RAHC*)	17, 19, 40
Chlordiazepoxide hydrochloride	7, 17
Chlordimeform (Galecron*)	7, 17
Chlorendic acid (*RAHC*)	17, 40
Chlorethoxyfos	17, 32
Chlorfenac (Fenatrol*)	3, 17
Chlorfenapyr	17, 26

Name(s)	RGN
Chloroform (Suspected Carcinogen; *RAHC*)	17, 40
Chloroformates, n.o.s. (All forms)	13, 17
Chloroformoxime	17, 66
Chlorofos*	17, 32
Chloroheptane(s)	17
Chlorohexane(s) (Dichlorohexane, Trichlorohexane, etc.)	17
Chlorohydrin	4, 17
Chloro-IPC (Isopropyl *n*-[3-chlorophenyl]carbamate, CIPC)	9, 17
Chloro-*N*-isopropylacetanilide	6, 17
Chloromercuriferrocene	24
Chloromethane	17
Chloromethylaniline (Various forms)	7, 17
Chloromethylchloroformate	13, 17, 67
Chloromethylchlorosulfonate	17
Chloromethyl ether as Bis(chloromethyl) ether (Human carcinogen)	14, 17, 40
Chloromethyl ethyl ether (Suspected Human carcinogen)	14, 17, 35
Chloromethyl methyl ether (Human carcinogen)	14, 17, 35, 40
Chloromethylnaphthalene (Lachrymator)	17
Chloromethylphenol	17, 31
Chloromethylphenoxyacetic acid (MCPA)	3, 17
3-Chloro-4-methylphenyl isocyanate	17, 18
Chloromethylphosphonic dichloride	17
3-Chloro-2-methylpropene (*RAHC*)	17, 40
Chloronaphthalene oil	17
Chloroneb	17
Chloro-*m*-nitroacetophenone	17, 27
Chloronitroaniline(s)	17, 27
Chloronitrobenzene (*ortho*-, *meta*-)	17, 27
p-Chloronitrobenzene (Animal carcinogen)	17, 27

Name(s)	RGN
Chloronitrobenzenesulfonic acid (Various forms) 3, 17, 27	
Chloronitrobenzoic acid 3, 17, 27	
Chloronitrobenzotrifluoride 17, 27	
Chloronitropropane (Korax*) 17, 27, 62	
Chloronitrotoluene(s) 17, 27	
Chloropentafluoroacetone 17, 39, 67	
Chloropentafluoroethane (Monochloropentafluoroethane) 17, 39	
Chloropentane 17	
Chloroperoxybenzoic acid 17, 30	
Chlorophacinone 17, 19	
Chlorophenamidine 7, 17	
Chlorophenate(s) (e.g. sodium chlorophenate) 17	
Chlorophenolate(s) 17, 31	
Chlorophenol(s) 17, 31	
Chlorophenoxyacetic acid 3, 17	
Chlorophenoxy herbicide(s) (2,4-D; 2,4-DB; Dichlorprop; Erbon*; Falone*; MCPA; MCPB; Mecoprop; Silvex; 2,4,5-T) 3, 17	
Chlorophenoxypropionic acid (Chlorprop) 3, 17	
4-Chloro-o-phenylenediamine (*RAHC*) 7, 17, 40	
Chlorophenyl isocyanate(s) 17, 18	
2-Chlorophenyl methylcarbamate 9, 17	
3-(*p*-Chlorophenyl)-5-methylrhodanine 17, 20	
Chlorophenylphenol (Various forms) 17, 31	
Chlorophenyl phenylsulfone (Chlorodiphenyl sulfone) 17	
Chlorophenyltrichlorosilane 17, 67	
Chlorophos 17, 32	
Chlorophthalic acid 3, 17	
Chloropicrin (Chlorpicrin; Lachrymator) 17, 27	
Chloropivaloyl chloride 17, 67	
Chloroplatinic acid 3, 17, 24	

Name(s)	RGN
Chlorotrifluoroethylene	17, 35, 67
Chlorotrifluoromethane	17, 39
Chlorotrifluoromethylaniline	7, 17
Chlorotrinitrobenzene	17, 27, 62
Chlorovinyldichloroarsine	17, 24, 67
Chlorovinylmethylchloroarsine (Vesicant gas; Poison)	17, 24, 67
para-Chloro-*meta*-xylenol	17, 31
Chlorox (Sodium/potassium hypochlorite in basic solution)	10, 64
Chloroxuron (Tenoran*)	6, 17
Chloroxylenol (Various forms)	17, 31
6-Chloro-3,4-xylyl methylcarbamate	9, 17
Chloroxynil	17, 26
Chlorpicrin	17, 27
Chlorpromazine	7, 17
Chlorpyrifos (Chlorpyriphos, Dursban*)	17, 32
Chlorquinox	17
Chlorthion	17, 27, 32
Chlorthiophos	17, 32
Chlozolinate	13, 17
Cholecalciferol	61
Chromated copper arsenate (CCA)	24
Chrome alum (Chrome potash alum)	24
Chrome pigment	24
Chromic acetate	24
Chromic acid (Human carcinogen; Poison)	3, 24, 40, 64
Chromic anyhdride	24, 64, 67
Chromic chloride	24
Chromic fluoride	15, 24
Chromic nitrate	24, 62, 64
Chromic oxide	24
Chromic sulfate (Chromic sulphate)	24
Chromic sulfide	24, 33

Name(s)	RGN
Clethodim	4, 19, 20
Cleve's acid	3
Clopidol	7, 17
Cloprop (Fruitone* CPA)	3, 17
Clopyralid	3, 17
CMME (Chloromethyl methyl ether; Human carcinogen)	14, 17, 35, 40
CMPP	3, 17
Coal gas	36
Coal tar pitch volatiles (Human carcinogen)	31, 40
Cobalt (Dust; Flammable; Animal carcinogen)	22, 66
Cobalt-57,58,60 (Radioactive poisons, Compounds & Element)	23
Cobalt acetate (Cobaltous acetate)	24
Cobalt arsenate (Cobaltous arsenate)	24
Cobalt bromide	24
Cobalt carbonyl	24
Cobalt chloride	24
Cobalt chromate (Cobaltous chromate; carcinogen)	24
Cobalt compounds, inorganic	24
Cobalt hydrocarbonyl, gas	36
Cobaltite dust	24
Cobalt naphthenate	24
Cobalt nitrate (Cobaltous nitrate)	24, 64
Cobaltocene	24
Cobaltous bromide (Cobalt bromide)	24
Cobaltous chloride (Cobalt chloride)	24
Cobaltous chromate (Suspected carcinogen)	24
Cobaltous cyanide	11, 24
Cobaltous fluoride (Cobalt difluoride)	15, 24
Cobaltous formate	24
Cobaltous nitrate	24, 64
Cobaltous perchlorate	24, 64
Cobaltous resinate (Cobalt resinate)	24, 66

Name(s)	RGN
Cuprous acetylide...............................	24, 62, 65, 67
Cuprous cyanide................................	11, 24
Cuprous iodide (Fume, Dust, or Mist).....................	24
Cuprous mercuric iodide.............................	24
Cuprous potassium cyanide	11, 24
Cuprous sulfide.................................	33
Curalin*	17, 19
Curasol*	9
Curium (Synthetic radioactive element).....................	23
Cutless*......................................	4, 17
Cyanamide	6
Cyanazine	17, 26
Cyanic acid	36, 62
Cyanides, inorganic, n.o.s.	11
Cyanoacetamide	6, 26
Cyanoacetic acid................................	3, 26
Cyanobrik*	11
Cyanochloropentane	17, 26
Cyanoethyl acrylate	13, 26
Cyanofenphos (Cyanophenphos)	26, 32
Cyanogas*.....................................	26, 67
Cyanogen (Gas or Liquid)	11, 26, 36
Cyanogen azide..................................	62, 66
Cyanogen bromide	11, 17
Cyanogen chloride (Gas or Liquid)	11, 17, 39
Cyanogen fluoride (Gas)	11, 17, 39
Cyanogen iodide.................................	11, 17
Cyanogran*....................................	11
Cyanomethyl acetate...............................	13, 26
Cyano(methylmercuri)guanidine or Cyano (methylmercury)guanidine............................	6, 24, 26
(S)-Cyano(3-phenoxyphenyl)methyl-(S)-chloro-*alpha*-(1-methylethyl)benzeneacetate	17, 26

D

Name(s)	RGN
DDT (Human carcinogen; *RAHC*)	17, 40
DDVP	17, 32
2,4-DEB	17
Decaborane	62, 66, 67
Decabromobiphenyl (*RAHC*)	17, 40
Decalin (Decahydronaphthalene)	29
Decamethrin	17, 26
Decanal	5
Decane	29
Decanol	4
Decanoyl peroxide	30
Decarbofuran	9
Decene	28
Decylamine	7
Decyl benzene	16
Deet (DEET*)	6
DEF 6* (Tribufos)	32
DEGN	27, 62
DEHP (*RAHC*)	13, 40
Dehydroacetic acid (DHA)	19
Deiquat	7, 17
Delnav*	32
Delsam*	17, 32
Deltamethrin	17, 26
Demephion-S (or Demephion-O)	32
Demeton (Systox)	32
Demeton-O-methyl	32
Demeton-S-methyl	32
DEN (*RAHC*)	7, 40
Denatured alcohol	4
2,4-DEP	17, 32
DES (Human carcinogen)	4, 40
2,4-DES	3, 17

E

Name(s)	RGN
1,2-Epoxy-3-ethoxypropane	34
Epoxyheptachlor (HCE; Carcinogen; Poison)	17, 34
1,2-Epoxypropane	34
2,3-Epoxy-1-propanol (Animal carcinogen)	4, 34
2,3-Epoxypropyl ether or Bis-(2,3-Epoxypropyl) ether	34
EPTC	12
Erbium, finely divided form	22
Erbium nitrate	62
Erbium oxalate	24
Erbon	13, 17
Erionite (Human carcinogen)	40
Erythrite	24
Erythritol anhydride (Butadiene dioxide)	3
Erythrityl tetranitrate (Erythrol tetranitrate)	62
Esbiol*	61
Esdepalléthrine	61
Esfenvalerate	17, 26
Esters, n.o.s.	13
Estox*	32
Estriol and Estradiol (Carcinogens)	4
Etaconazole	7, 17
Ethalfluralin	17, 27
Ethane	29, 35
Ethanedithiol	20
Ethanethiol	20
Ethanoic acid	3
Ethanol	4
Ethanolamine (MEA)	4, 7, 10
Ethanoyl chloride	17, 67
Ethazcate*	12
Ethchlorvynol	17, 28
Ethene	28, 35, 63
Ethephon (Ethrel*)	3, 17

Name(s)	RGN
Ethers, n.o.s.	14
Ethers (Containing peroxides)	62
Ethiofencarb	9, 17, 20
Ethiolate	12
Ethion	32
Ethofumesate	13
Ethoprop (Ethoprophos)	32
Ethoxycarbonyl isothiocyanate (Lachrymator)	18, 67
Ethoxydimethylsilane	24
Ethoxyethanol	4, 14
2-Ethoxyethyl acetate	13
2-[1-(Ethoxyimino)butyl]-5-[2-(ethylthio)propyl]- 3-hydroxy-2-cyclohexen-1-one	19, 20
Ethoxymethylenemalononitrile	26
Ethoxyquin (Under acidic conditions; Hay, Silage, etc.)	26
Ethrel*	3, 17
Ethyl abietate	13
N-Ethylacetanilide	6
Ethyl acetate	13, 62
Ethyl acetoacetate	13
Ethylacetylene	28, 35, 63
Ethyl acrylate (Suspected human carcinogen)	13, 62, 63
Ethyl alcohol	4
Ethyl allylacetoacetate (Various forms)	13
Ethylaluminum dichloride (EADC)	66, 67
Ethylaluminum sesquichloride (EASC)	66, 67
Ethylamine, gas or liquid	7, 10, 37
2-Ethylamino-4-isopropylamino-6-methylthio- s-triazine	7
Ethyl amyl ketone (EAK)	19
Ethylaniline (Various forms)	7
Ethyl anthranilate	7, 13
Ethylarsenious oxide	24

Name(s)	RGN
Ethylene glycol monomethyl ether acetate....................	13
Ethyleneimine (Ethylenimine; Carcinogen)...............	7, 62, 63
Ethylene oxide (Human carcinogen)	34, 36, 40, 63
Ethylene thiourea (*RAHC*)	9, 40
Ethyl ether (Diethyl ether)	14
Ethyl ether containing peroxides	14, 62
Ethyl fluoride	17, 35
Ethyl formate	13
Ethyl formylpropionate (Various forms)	13
Ethylfuran	14
Ethylhexaldehyde(s).....................................	5, 66
Ethylhexanediol	4
Ethylhexene (Various forms)	28
2-Ethylhexoic acid	3
2-Ethylhexyl acrylate	13, 63
2-Ethylhexylamine	7
N-2-Ethylhexylaniline................................	7
Ethylhexyl chloride....................................	17
2-Ethylhexyl chloroformate	13, 17
Ethylhexyl cyanoacetate	26
Ethylidene chloride....................................	17
Ethylidene fluoride....................................	17, 35
5-Ethylidene-2-norbornene (ENB)....................	28
Ethyl iodide..	17
Ethyl iodoacetate	17, 66
Ethyl isobutyrate.....................................	13
Ethyl isocyanate	18, 67
Ethyl lactate	13
Ethyllithium	65, 66, 67
Ethyl mercaptan	20
Ethylmercuric acetate	24
Ethylmercury bromide (Ethylmercuric bromide)............	24
Ethylmercury chloride (Ethylmercuric chloride)	24

Name(s)	RGN
Ethyl propyl ether	14
Ethyl propyl ketone	19
Ethyl silicate	24
Ethyl sulfide (Diethyl sulfide)	20
Ethylsulfuric acid (Ethylsulphuric acid)	3, 67
Ethylthiocarbimide (Ethyl mustard oil)	18
Ethylthiodemeton (M-74, Disulfoton)	20, 32
Ethyl thioethanol	4, 20
N-Ethyl toluidine(s)	7
Ethyltrichlorosilane	17, 66, 67
Ethyl tuex*	20
Ethyl vinyl ether	14
Ethylzinc	24
Ethyne (Gas)	28, 35
Etinofen	27, 31
Etoc*	13
Etridiazole	17, 20
Euparen* (Euparen* M, Tolylfluanid; Dichlofluarid)	6, 17
Europium, powder	21, 66, 67
Europium isotope (Radioactive)	23
Evisect*	3, 20
Exothion	32
Explosives, n.o.s.	62
Exxpel 2*	19, 31

F

Name(s)	RGN
Facet* (Fasnox*)	3, 17
F-acid (Casella's acid)	3
Famphur (Famophos)	32
Faneron*	17, 27
Fenac	3, 17

Name(s)	RGN
Folpet	6, 17
Fomesafen	17, 27
Fonofos* (Fonophos)	32
Fopel	17, 20
Formaldehyde (*RAHC*; Nasal cancer)	5, 40
Formalin (Formaldehyde with alcohol)	5
Formamide	6
Formetanate (Formetanate hydrochloride)	9
Formic acid	3, 65
Formonitrile	26
Formothion	32
Formyl fluoride	39, 67
Fortress*	17, 32
Fosfamid	32
Fosthiazate	32
Fostion*	32
FP acids*	3, 17
Freon* (Various forms)	17, 39
Frontier*	6, 17
Fthalide	17
Fuel oil (no. 1 through no. 6)	61
Fuels (Petroleum distillates, Gasoline, Jet fuel, etc.)	61
Full*	17, 26
Fumaric acid	3
Fumarin*	19
Fumaryl chloride	17, 67
Furadan* (Furadex*; Furacarb*)	9
Furaldehyde(s)	5, 63
Furan	14
Furazolidone (carcinogen)	7, 27
Furfural (Animal carcinogen)	5, 63
Furfuraldehyde (Animal carcinogen)	5, 63
Furfuran	6, 14

G

I

Name(s)	RGN
Isoamyldichloroarsine	17, 24, 67
Isoamyl formate	13
Isoamyl mercaptan	20
Isobenzan	17
Isobornyl thiocyanoacetate	18
Isobutane	29, 35
Isobutanol	4
Isobutene, liquid or gas	28, 35
Isobutyl acetate	13
Isobutyl acrylate	13, 63
Isobutyl alcohol	4
Isobutyl aldehyde	5
Isobutylamine	7
Isobutylbenzene	16
Isobutyl chloroformate	13, 17, 67
Isobutylene	28, 35
Isobutyl formate	13
Isobutyl isobutyrate (IBIB)	13
Isobutyl isocyanate	18, 67
Isobutyl mercaptan	20
Isobutyl methacrylate	13, 63
Isobutyl propionate	13
Isobutyraldehyde	5
Isobutyric acid	3
Isobutyric anhydride	3, 67
Isobutyronitrile	26
Isobutyryl chloride (Isobutyroyl chloride)	17
Isocil (Isoprocil)	6, 17
Isocyanate solution, n.o.s.	18, 67
Isocyanate(s), n.o.s.	18, 67
Isocyanatobenzotrifluoride	17, 18, 67
Isocyanic acid	36, 62
Isodecyl acrylate	13

J

K

M

Name(s)	RGN
Magnesium chromate	24
Magnesium diamide	67
Magnesium diphenyl	67
Magnesium fluoride	15
Magnesium fluosilicate (Magnesium fluorosilicate)	15
Magnesium granules, coated	21
Magnesium hydride	67
Magnesium hydroxide	10
Magnesium nitrate	64
Magnesium oxide, fume	24
Magnesium perborate	64
Magnesium perchlorate	64
Magnesium permanganate	64
Magnesium peroxide	64
Magnesium phosphide	67
Magnesium powder	21
Magnesium silicide	67
Magnesium silicofluoride	15
Magnesium sulfide	33, 65
Malachite Green (Victoria Green)	17
Malathion (Malatop*, Maldison)	13, 32
Maleic acid	3, 67
Maleic anhydride	3, 67
Maleic hydrazide	6
Malic acid (Hydroxysuccinic acid, Not Maleic acid)	3
Malonamide	6
Malonic acid	3
Malonic dinitrile	26
Malononitrile	26
Mancozeb (Mancozebe, Mancozin*)	12
Mandelic acid	3
Mandelonitrile (Laetrile*)	26
Maneb (Manex*)	12

Name(s)	RGN
Mercury fulminate (Mercuric cyanate)	24, 62
Mercury gluconate	24
Mercury iodide	24
Mercury metal	22, 24
Mercury nitride (Trimercury dinitride)	24, 25, 62
Mercury nucleate	24
Mercury oleate	24
Mercury oxide	24
Mercury oxycyanide, desensitized	11, 24
Mercury potassium iodide	24
Mercury salicylate	24
Mercury selenide	24
Mercury sulfate (Mercury sulphate)	24
Mercury thiocyanate	24
Mercusol	24
Mescaline	7
Mesitylene	16
2-Mesitylenesulfonyl chloride	17
Mesityl oxide	19
Mesurol*	9, 20
Metacid* TS	12
Metafos	27, 32
Metal alkyl solution, n.o.s.	24
Metal alkyl halides, n.o.s.	24, 67
Metal alkyls, n.o.s.	24, 67
Metal aryl halides, n.o.s.	24, 67
Metal aryl hydrides, n.o.s.	24, 67
Metal aryls, n.o.s.	24, 67
Metalaxyl	13
Metal carbonyls, n.o.s.	24
Metal catalyst, dry (Water and/or air sensitive)	21
Metal compounds, n.o.s. (Toxic)	24
Metaldehyde	5

Name(s)	RGN
MXDA	7
Myclobutanil	7, 17, 26
Mycotox*	17
Myleran®* (Human carcinogen)	4, 40
Myristoyl peroxide	30

N

Name(s)	RGN
Nabam	12
Nac*	9
NaK (Sodium-potassium alloy)	21
Naled	17, 32
Napalm	3, 24
Naphite	27, 62
Naphtha (Benzin)	61
Naphthacene	62
Naphthalene	16
Naphthaleneacetamide	6, 16
Naphthaleneacetic acid	3, 16
Naphthenic acid	3
Naphthol	16, 31
Naphthoxyacetic acid	3
Naphthylamine (Several forms; Carcinogen)	7
2-Naphthylamine (β-Naphthylamine; Human carcinogen)	7, 40
Naphthyl mercaptan	20
1-Naphthyl-*N*-methylcarbamate	9
Naphthylphthalamic acid	3, 6
Naphthylthiourea (ANTU)	9
Naphthylurea	6
Naramycin	19
Natural gas	35
Natural gasoline	61

O

Name(s)	RGN
Orthoboric acid	3
Oryzalin	27
Osmium	22
Osmium amine nitrate	24, 64
Osmium amine perchlorate	24, 64
Osmium tetroxide (Osmic acid)	24
Oxadiazon	17, 19
Oxalates, metal, water soluble	24
Oxalic acid	3
Oxalyl chloride	17, 67
Oxamyl	9, 20
Oxidizing Agents/Oxidizers, strong, n.o.s.	64
Oxone*	64
Oxsol 100*	17
p,p'-Oxybis(benzenesulfonyl hydrazide)	7
Oxydemeton-methyl	32
4,4'-Oxydianiline (*RAHC*)	7, 40
Oxyfluorfen	17, 27
Oxygen	38, 64
Oxygen-17,18 (Radioactive element, Heavy oxygen)	23, 64
Oxygenated Hydrocarbon(s)* (Organic acids & esters)	3, 13
Oxygen difluoride (Oxygen fluoride)	17, 38, 64, 66, 67
Oxythioquinox	20
Ozone	38, 64

P

Name(s)	RGN
PABA	3, 7
Padan	12, 17
PAHs (Polycyclic aromatic hydrocarbons; *RAHC*)	16, 40
Palladium nitrate	24, 64

Name(s)	RGN
Performic acid	3, 62, 64
Periodic acid	2, 64
Perlite (Nuisance dust)	24
Permanganate, n.o.s.	64
Permanganates, inorganic, n.o.s.	64
Permethrin (Penncapthrin*)	13, 17
Permonosulfuric acid	2
Peroxides, inorganic, n.o.s.	64
Peroxides, organic, n.o.s.	30
Peroxyacetic acid	3, 30
Peroxysulfuric acid	2, 64
Persulfates (Persulphates), inorganic, n.o.s.	64
Perthane*	17
PETN	27, 62
Petrol	61
Petroleum crude oil and distillates	61
Petroleum distillates (Paints, Oils, Lacquer, Paint thinners, Grease, etc.)	61
Petroleum ether	61
Petroleum gas (Liquefied, Compressed, LPG)	35
Petroleum hydrocarbons (Gasoline, Fuel oil, Kerosene, Turpentine, Motor oils, etc.)	61
Petroleum naphtha	61
Petroleum spirits (Naphtha, Spirits, Petroleum ether)	61
Pharaoh's serpent eggs (Mercuric thiocyanate)	18, 24
alpha-Phellandrene	28
Phenacaine hydrochloride	6
Phenacetin (RAHC)	6, 40
Phenacyl bromide	17, 19
Phenacyl chloride	17, 19
Phenacyl fluoride	17, 19
Phenanthrene (Carcinogen)	16
Phenarsazine chloride	7, 17, 24

GUIDE TO CHEMICAL REACTIVITY AND COMPATIBILITY **145**

GUIDE TO CHEMICAL REACTIVITY AND COMPATIBILITY

Name(s)	RGN
Phosphorus acid (Phosphonic acid, ortho)	2, 66, 67
Phosphorus heptasulfide	33, 65, 67
Phosphorus oxybromide	17, 64, 67
Phosphorus oxychloride	17, 64, 67
Phosphorus pentabromide (Phosphoric bromide)	17, 67
Phosphorus pentachloride (Phosphoric chloride)	17, 67
Phosphorus pentafluoride	17, 39, 66, 67
Phosphorus pentasulfide	33, 65, 66, 67
Phosphorus pentoxide	67
Phosphorus sesquisulfide	33, 65, 67
Phosphorus tribromide	17, 67
Phosphorus trichloride	17, 67
Phosphorus triiodide	17, 67
Phosphorus trisulfide	33, 65, 67
Phosphoryl bromide	17, 64, 67
Phosphoryl chloride	17, 64, 67
Phosphothiate(s), n.o.s.	32
Phosphotungstic acid	2, 64
Phoxim (Phoxime)	26, 32
Phtalofos	32
Phthalic acid	3
Phthalic anhydride	3, 67
Phthalide	17
Phthalimide derivative pesticide	6
m-Phthalodinitrile (IPM)	26
Phthalonitrile	26
Phthaloyl chloride	17, 67
Phytex*	12
Picloram (Piclorame)	3, 17
Picoline(s) (Methylpyridines)	7
4-Picolylamine	7, 10
Picramic acid (Picraminic acid)	7, 27, 62
Picramide	7, 27, 62

Name(s)	RGN
Pyranica (Tebufenpyrol)	7, 17
Pyrazole	7
Pyrazoline	7
Pyrazophos	32
Pyrazoxyfen	7, 17, 19
Pyrellin* (Pyrethrins plus Rotenone)	13, 19
Pyrene (Carcinogen)	28
Pyrethrins I (Pyrethrines)	13, 19, 66
Pyrethrins II (Pyrethrines)	13, 19
Pyrethroid pesticides	13, 19
Pyrethrum	13, 19
Pyridaben	17, 19, 20
Pyridaphenthion	32
Pyridate	17, 20
Pyridine	7
Pyrifenox	7, 17
Pyrimiphos-ethyl	32
Pyrimithate	32
Pyrinox*	17, 32
Pyrocatechol (Animal carcinogen)	4
Pyrogallol	31
Pyrolan	9
Pyromellitic acid (PMA)	3
Pyromellitic dianhydride (PMDA)	3, 67
Pyroquilon	19
Pyrosulfuric acid	2
Pyrosulfuryl chloride (Pyrosulphuryl chloride)	17, 67
Pyroxylin	27, 62
Pyrrole	7
Pyrrolidine	7
Pyruvic acid	3
Pyruvic alcohol	4, 19

Q

Name(s)	RGN
Quicksilver	22, 24
Quinaldine	7
Quinalphos	32
Quinclorac	3, 17
Quinhydrone	4
Quinine	7, 28
Quinoclamine	7, 17, 19
Quinoline	61
Quinone	19
Quinsol (Quinosol)	24
Quintozene	17, 27
Quizalofop-ethyl or Quizalofop-P-ethyl	13, 17

R

Name(s)	RGN
Rabcide*	17
R-acid	3
Radiant*	6, 17
Radium-226 (Radioactive compounds)	23
Radium-226 (Radioactive toxic element, Destructive)	23, 66, 67
Radon-222 (Radioactive toxic gas; Human carcinogen)	1, 23, 40
Ramrod*	6, 17
Raney nickel	21
Rare gas	1
Raticate	6
Raxil*T	12
RDX	27, 62
Red arsenic	24, 33, 65
Red lead	24, 64
Red phosphorus	21, 65, 66
Reducing agents, strong, n.o.s.	65

T

U

V

W

X

Name(s)	RGN
Xylene(s)	16
m-Xylene-α,α'-diamine (MXDA)	7
Xylenol(s) (2,4-Xylenol, Dimethylphenol, etc.)	31
Xylidine(s) (Animal carcinogen)	7
Xylycarb	9
Xylyl bromide	17
Xylyl chloride	17
Xylyl methylcarbamate	9

Y

Name(s)	RGN
Yellow AB	7
Yellow OB	7
Yellow phosphorus	65, 66
Yellow salt	23, 24, 64
Ytterbium fluoride	15
Yttrium (Finely divided form)	22
Yttrium arsenide	24
Yukamate*	9

Z

Name(s)	RGN
Zeatin	7
Zeidane (DDT; Human carcinogen; *RAHC*)	17, 40
Zepar* BP	65
Zerlate*	12
Zeta -cypermethrin	17, 26
Ziman*	12
Zimate*	12
Zinc (Dust, powder, dry or damp)	21

Name(s)	RGN
Zinc (Foil)	21
Zinc-65 (radioactive poison)	23
Zinc acetylide	24, 65, 67
Zinc ammonium chloride	24
Zinc ammonium nitrite	24, 64
Zinc antimonide	24, 67
Zinc arsenate	24
Zinc arsenite (ZMA)	24
Zinc bisulfite (Zinc bisulphite)	24
Zinc borate	24
Zinc bromate	64
Zinc bromide	24
Zinc cadmium sulfide	24, 33
Zinc caprylate (zinc octanoate; decomposes in moist air)	66
Zinc carbonate	24
Zinc chlorate	64
Zinc chloride	24
Zinc chloride chromated	24
Zinc chromate (Human carcinogen)	24, 40
Zinc cyanide	11, 24
Zinc dichromate	24
Zinc diethyldithiocarbamate	12, 24
Zinc dimethyldithiocarbamate	12, 24
Zinc dimethyldithiocarbamate cyclohexylamine complex	7, 12
Zinc dioxide	24, 64, 67
Zinc dithionite (Zinc hydrosulfite)	24, 65
Zinc dithionite, aqueous (Zinc hydrosulfite, aqueous)	24, 65
Zinc dross	21, 67
Zinc ethyl (Diethylzinc; Pyrophoric)	24, 66, 67
Zinc fluorarsenate	15, 24
Zinc fluoride	15, 24

NOTE: An asterisk(*) or trade mark (®) after a chemical name identifies the trade name for the compound.

Name(s)   **RGN**

CHEMICAL REACTIVITY AND COMPATIBILITY CHART

1) The blanks in the chart indicate that the reactants in the two groups are normally unreactive, however the groups may be reactive if one of the reacting components contains another chemical substance *or* the conditions of the reaction change *or* extreme conditions exist.

2) A confined gas may produce an explosion *not* identified in this chart.

3) Some components in a destruction process (fire, explosion, etc.) or chemical reaction will decompose or be modified and then react with water, air or other chemicals in the mix to produce end products not expected in a chemical reaction.

4) The presence of air (excess oxygen), water (moisture), or a contaminant in a mixture can result in an explosion, fire, or an event which *cannot* be predicted at this time.

5) There will probably be a specific event that will result in a disaster or accident not identified or properly characterized in this table.

6) The *symbols in a parenthesis* indicate that some of the chemicals in a particular group will react (interact) with some of the chemicals of the second group.

7) Groups (RGNs) 1, 35 through 39 are in the gaseous phase and are represented by *italics*. These gaseous compounds will react (interact) with the gaseous phase of other compounds. Heated or evaporated solvents and/or solids will usually interact with gases. Often, solids and liquids decompose in a fire to form additional substances which will react with the gases.

8) Definitions: E = explosion; F = flammable/fire; G = gas emitted; H = heat given off; T = toxic gas emitted; etc. for P, V, S, and U.

	2	3	4	5	6	7	8	9
1	*(T)*							
2		HG	HG(E)	HG(T)	HT	HT	GTF(E)	H(T)
3	HG		H(P)	HP(T)		H(P)	HG(V)	
4	HG(E)	H(P)					G(H)	
5	HG(T)	HP(T)				H	GH	
6	HT							
7	HT	H(P)		H				
8	GTF(E)	HG(V)	G(H)	GH				
9	H(T)							
10	H(F)E	H(T)		H(P)				HG
11	T(F)	T					(G)	
12	HT(F)	TH		(T)H		HU	HU	
13	H(F)						(H)	
14	HF(E)							
15	HT	(T)						
16	(H)(F)							
17	HT					H(T)	G(H)	
18	HGT(F)	HG	(P)H			(H)	(G)	
19	HG(F)						(G)	
20	TH(F)	(T)					(G)	
21	FEH	HF(E)	FH(T)	(F)H	H(F)	H(T)F	HG(F)	FT
22	(E)(F)(T)	(F)T(E)	(T)(F)(H)				F(T)(P)	(F)
23	(H)(T)						(F)	
24	ST	ST	(S)T		(S)	(S)T		
25	HT(E)	HT(E)	TH(E)	H(P)T			U	HF
26	HT(E)	H(T)						
27	HT(E)			H		T(E)(F)	*(E)*	
28	HF(E)			H				
29	(F)(E)					(E)(F)		
30	H(V)(E)		HF	VG(E)		HT(F)(P)	ETV	HT
31	H						HG	
32	HT							
33	THF	HT		(H)			(E)T	
34	H(P)(F)	H(P)	H(P)			H(T)P	GH	
35	*(F)(E)(T)*	*(F)(T)*	*(F)(E)(T)*	*(F)*	*F(T)*	FT	*(F)(E)T*	*(F)(T)*
36	*HTF(E)*	*(F)(T)*	*(F)(T)(E)*	*(F)(T)*	*(F)(T)*	*(T)*	*(F)T(E)*	*(F)(T)*
37	FTE	TF	TFE	*(F)(T)*		*(T)*	FTE	*(F)(T)*
38	*(T)(E)(F)*	*(T)(F)(E)*	*(H)(T)FE*	H(E)(F)	*HT(E)*	FT	*HT(E)(F)*	*HF(T)*
39	*(T)(F)*	*(T)(F)*	*(T)(F)*			FT	FT	*(F)(T)*
40	T(F)(E)S	T(S)	(S)(T)	(S)(T)	(T)	(T)	(T)(H)	(F)(T)
61	(F)(T)(E)	(T)				(E)(F)		
62	(T)(F)(E)	(E)T(F)	(F)(E)			T(E)(F)	EV(F)	(E)(H)T
63	HP(T)	(P)					(P)	
64	FT(V)(E)	HG(T)	H(E)F(T)	H(E)(F)	HT(E)	HT(E)	HT(E)(F)	HT
65	(F)(E)(T)	H(F)	F(E)	H(F)	H(F)	H(F)T	H(G)	(H)(F)T
66	——	——	REACTS	IN	AIR	——	REACTS	IN
67	(T)(F)(E)	(T)	E(T)F	(E)T(F)	T	T	ETF	T

10	11	12	13	14	15	16	17	
								1
H(F)E	T(F)	HT(F)	H(F)	HF(E)	HT	(H)(F)	HT	2
H(T)	T	HT			(T)			3
								4
H(P)		(T)H						5
								6
		HU					H(T)	7
	(G)	HU	(H)				G(H)	8
HG								9
			(H)				H(F)	10
							(H)	11
								12
(H)								13
								14
								15
								16
H(F)	(H)							17
HF(G)	HG	U						18
(H)	(T)							19
							H	20
H(F)	H(F)	H(T)	HF	HF(E)			(E)F(T)	21
HF(E)							(E)(F)	22
(H)(F)							(H)(F)	23
(S)T	T							24
	(F)		H(F)	(F)(E)			HT	25
								26
(E)				(F)(E)				27
						S		28
						S		29
	ET	HT		(E)FH		(E)(F)H	(E)(T)	30
								31
H(E)								32
								33
HP	H(P)			S				34
	F	FT	*(T)(F)*	*(F)(E)*			(F)(T)	35
(H)(T)	(T)	(T)	*(T)(F)*	*(F)(E)*			(F)(T)	36
			(T)(F)	*(F)(E)*			(F)(T)	37
(T)		HT	*(F)HT*	EFH		(E)(F)H	(H)(T)	38
(T)				*(F)(T)*				39
(S)(T)	(T)	(T)	(T)	(T)	(T)	(S)(T)	(S)(T)	40
(F)(T)				(E)(F)		(S)		61
H(E)(F)			ET	(E)(F)		(E)(F)		62
HP(T)								63
(T)	(T)	HT(E)	H(E)	HEF		(E)FH	HT(F)	64
		(T)	FH	(F)T(E)			H(E)(T)	65
AIR	—	—	—	—	—	—	REACTS	66
H(E)T(F)	(F)T	(E)T(F)	(F)(T)	ETF	(T)		(E)TF	67

	18	19	20	21	22	23	24	25
1								
2	HGT(F)	HG(F)	TH(F)	FHE	(E)(F)(T)	(H)(T)	ST	HT(E)
3	HG		(T)	FH(E)	(F)T(E)		ST	HT(E)
4	(P)H			FH(T)	(T)(F)(H)		(S)T	HT(E)
5				(F)H				H(P)T
6				(F)H			(S)	
7	(H)			FH(T)			(S)T	
8	(G)	(G)	(G)	(F)HG	F(T)(P)	(F)		U
9				FT	(F)			HF
10	HF(G)	(H)		(F)H	HF(E)	(H)(F)	(S)T	
11	HG	(T)		(F)H			T	(F)
12	U			H(T)				
13				HF				H(F)
14				HF(E)				(F)(E)
15								
16								
17			H	(T)F(E)	(E)(F)	(H)(F)		HT
18			(H)(P)	F(E)H	(F)			U
19			(H)	FH				FH
20	(H)(P)	(H)		(F)HT	(H)			FH
21	F(E)H	FH	(F)HT			(F)(T)	(T)(F)	F(T)E
22	(F)		(H)					
23				(F)(T)				
24				(T)(F)				
25	U	FH	FH	EF(T)				
26				FH(P)				
27				(V)(E)				FV(E)
28				F	(F)		(F)	
29								
30	GH	(V)	(T)HE	EV	H(E)(F)	(F)(T)	H(G)	TF(E)
31	H(T)			FH				FH
32				H(E)				
33	HG(T)							
34			(P)(T)	H(E)F	H(P)		(H)T	(H)(P)
35	*(F)(T)*	*(F)(T)*	*FT(E)*	*F(T)(E)*			*(T)*	*(F)(E)T*
36	*(F)(T)*	*(F)(T)*	*FT(E)*	*FT(E)*	*(T)*		*(T)*	*(F)(E)T*
37	*(F)(T)*	*(F)(T)*	FT(E)	FT(E)			*(T)*	*F(E)T*
38	*(F)(T)(E)*	*F(T)(E)*	*FT(E)*	*(E)(V)*	*(T)*	*(F)*	*(T)*	*T(F)HV*
39	*(F)(T)*	*F(T)*	*FT*	*(F)(T)*			*(T)*	*(F)(T)*
40	(T)	(S)(T)	(T)	T(F)(E)	(T)	(T)	(T)	(T)(E)
61				FT(E)(G)	(T)(F)(E)			TF
62	(T)(F)(E)	(H)(T)	T(F)(E)	EVF	(E)(T)	(T)(S)	(T)(E)	V(E)FT
63	(T)(P)			(P)(F)	(T)(F)		(P)(T)	(P)FHT
64	H(F)T(P)	HF	T(F)(E)	HFE	H(F)(E)	(T)(S)	(H)T	VT(E)F
65	H(G)(F)T	H(G)F	(F)HT	(E)T(F)	(T)		(H)T(F)	(H)T(F)
66	IN	AIR	——	REACTS	IN	AIR	——	——
67	(E)TF	(E)T(F)	(E)TF	ETF	(E)(T)F	(T)	(E)T(F)	(E)TF

198

26	27	28	29	30	31	32	33	
				(T)				1
HT(E)	(E)HT	HF(E)	(F)(E)	H(V)(E)	H	HT	THF	2
H(T)							HT	3
				HF				4
	H	H		VG(E)			(H)	5
								6
	T(E)(F)	(E)(F)		HT(P)(F)				7
	(E)			ETV	HG		(E)T	8
				HT				9
	(E)					H(E)		10
				ET				11
				HT				12
								13
	(F)(E)			(E)FH				14
								15
		S	S	(E)(F)H				16
				(T)(E)				17
				GH	H(T)		HG(T)	18
				(V)				19
				(T)HE				20
FH(P)	(V)(E)	F		EV	FH	H(E)		21
		(F)		H (E)(F)				22
				(F)(T)				23
		(F)		H(G)				24
	FV(E)			TF(E)	FH			25
				(T)(P)				26
				(E)(F)T				27
				HP				28
								29
(T)(P)	(E)(F)T	HP			H(T)	(T)(F)	TGH	30
				H(T)				31
				(T)(F)				32
				TGH				33
				(H)(F)(E)			(T)	34
(T)(F)	FT(E)	(F)(T)	(F)(T)	*(T)(E)(F)*	(F)	*(T)*		35
(T)(F)	(F)T(E)	(F)(T)	(F)(T)	*(T)(E)(F)*	(F)	*(T)*		36
(T)(F)	FT(E)	(F)(T)	(F)(T)	FET	(F)(T)	*(T)*		37
F	FT(E)	(T)(F)	(T)(F)		(E)(F)	*(T)*	*(T)*	38
(T)(F)	(F)(T)			*(T)(F)*	(F)(T)			39
(T)	(T)	(T)(S)	(T)(S)	(T)	(T)	(T)	(T)	40
	T(F)(E)	S	S	F(E)T	(T)		T(E)	61
T(F)(E)	(E)(T)	F(E)	F(E)TV	(E)(V)F	(F)(E)	T(F)	(E)TF	62
	(F)			TH(F)(E)	H(P)T		T(P)(E)	63
(T)(F)(E)	(E)VF	HF(E)	H(F)(E)(V)	(H)(F)(E)	TF(E)	T(H)(F)	HTF(E)	64
H(F)(T)	H(F)(E)	(F)(E)		EVFHT	(F)H	(T)(F)	T(E)(F)	65
——	——	——	——	REACTS	IN	AIR	——	66
(T)(F)	(E)TF	T(E)F	(E)TF	(E)TF	(E)TF	TF	(E)TF	67

	34	35	36	37	38	39	40	61
1					(T)			
2	H(P)(F)	(F)(E)(T)	F(E)TH	FTE	(F)(E)(T)	(F)(T)	T(F)(E)S	(T)(F)(E)
3	H(P)	(F)(T)	(F)(T)	FT	(F)(E)(T)	(F)(T)	T(S)	(T)
4	H(P)	(F)(E)(T)	(F)(E)(T)	FTE	(H)(T)FE	(F)(T)	(S)(T)	
5		(F)	(F)(T)	(F)(T)	H(E)(F)		(S)(T)	
6		F(T)	(F)(T)		HT(E)		(T)	
7	HP(T)	FT	(T)	(T)	TF	TF	(T)	(E)(F)
8	GH	(F)(E)T	(F)(E)T	FTE	HT(F)(E)	TF	(T)(H)	
9		(F)(T)	(F)(T)	(F)(T)	H(T)F	(T)(F)	(F)(T)	
10	HP		(H)(T)		(T)	(T)	(S)(T)	(F)(T)
11	H(P)	F	(T)				(T)	
12		FT	(T)		HT		(T)	
13		(F)(T)	(F)(T)	(F)(T)	(F)HT		(T)	
14	S	(F)(E)	(F)(E)	(F)(E)	FEH	(F)(T)	(T)	(E)(F)
15							(T)	
16					(F)(E)H		(S)(T)	(S)
17		(F)(T)	(F)(T)	(F)(T)	(H)(T)		(S)(T)	
18		(F)(T)	(F)(T)	(F)(T)	(F)(E)(T)	(F)(T)	(T)	
19		(F)(T)	(F)(T)	(F)(T)	F(E)(T)	F(T)	(S)(T)	
20	(P)(T)	F(E)T	F(E)T	FT(E)	F(E)T	FT	(T)	
21	H(E)F	F(T)(E)	FT(E)	FT(E)	(E)(V)	(F)(T)	T(F)(E)	TF(E)(G)
22	H(P)		(T)		(T)		(T)	(T)(F)(E)
23					(F)		(T)	
24	(H)T	(T)	(T)	(T)	(T)	(T)	(T)	
25	(H)(P)	(F)T(E)	(F)T(E)	FT(E)	VTH(F)	(F)(T)	(T)(E)	TF
26		(F)(T)	(F)(T)	F(T)	F	(F)(T)	(T)	
27		FT(E)	(F)T(E)	FT(E)	FT(E)	(F)(T)	(T)	(F)T(E)
28		(F)(T)	(F)(T)	(F)(T)	(F)(T)		(T)(S)	S
29		(F)(T)	(F)(T)	(F)(T)	(T)(F)		(T)(S)	S
30	(E)(F)(H)	(E)(F)(T)	(E)(F)(T)	EFT		(T)(F)	(T)	F(E)T
31		(F)	(F)	(F)(T)	(E)(F)	(T)(F)	(T)	(T)
32		(T)	(T)	(T)	(T)		(T)	
33	(T)				(T)		(T)	T(E)
34		(E)(F)T	(E)T(F)	(E)T(F)	(F)(H)	(T)(F)	(T)	(T)(E)(F)
35	(F)(E)T		(E)T(F)	(E)T(F)	(F)(E)T	(T)(F)	(T)	(F)(E)(T)
36	(F)(E)T	(E)T(F)		(E)T(F)	(E)T(F)	(T)(F)	(F)(T)(E)	(F)(E)T
37	(F)(E)T	(F)(E)T	(F)(E)T		(E)TF	(T)(F)	(T)	(F)T(E)
38	(F)(H)	(F)(E)T	(F)(E)T	(E)TF		(T)(F)	(T)	(F)T(E)
39	(F)(T)	(F)(T)	(F)(T)	(T)(F)	(T)(F)		(T)	(F)(T)
40	(T)	(T)	(F)(E)(T)	(T)	(T)	(T)		(F)(T)(S)
61	(T)(E)(F)	(F)(E)(T)	(F)(E)T	(E)T(F)	(F)T(E)	(F)(T)	(F)(T)(S)	
62	HF(E)T	(F)(T)(E)	(F)(E)T	(E)T(F)	(E)(F)T	T(F)(E)	(T)(E)(F)	(T)(E)(F)
63	(P)	(F)(T)	(F)T	T(F)	(F)T(E)	(T)	(T)(F)	(T)(F)(E)
64	HF(E)	(E)(F)(T)	(F)(E)T	(E)T(F)	T	T(F)	(T)(E)(F)H	F(E)TH
65	(F)(T)(H)	(E)(T)(F)	(F)(E)T	T(F)(E)	(E)T(V)F	(T)(F)	T(F)H(E)	(F)(T)(E)
66	——	——	REACTS	IN	AIR	——	——	REACTS
67	(E)TF(V)	(E)T(F)	(F)(E)T	(E)T(F)	(E)TF	T(F)(V)	(T)(F)(E)	(P)T(F)(E)

62	63	64	65	66	67	
		(T)				1
(T)(F)(E)	HP(T)	FT(V)(E)	(F)(T)(E)		(E)(T)(F)	2
(E)T(F)	(P)	HG(T)	H(F)	A	(T)	3
(F)(E)		HF(E)(T)	F(E)	I	E(T)F	4
		H(F)(E)	H(F)	R	(E)T(F)	5
		HT(E)	H(F)		T	6
T(E)(F)		HT(E)	H(F)T		T	7
EV(F)	(P)	HT(E)(F)	H(G)	S	ETF	8
(E)(H)T		HT	(H)(F)T	E	T	9
H(E)(F)	H(T)P	(T)		N	H(E)T(F)	10
		(T)		S	T(F)	11
		(H)(T)(E)	(T)	I	(E)T(F)	12
ET		H(E)	FH	T	(F)(T)	13
(E)(F)		HFE	(F)T(E)	I	ETF	14
				V	(T)	15
(E)(F)		HF(E)		E		16
		HT(F)	H(E)(T)		(E)TF	17
(T)(F)(E)	(T)(P)	H(F)T(P)	(F)H(G)T		(E)TF	18
(H)(T)		HF	FH(G)	A	(E)T(F)	19
T(F)(E)		T(F)(E)	(F)HT	N	(E)TF	20
EVF	(P)(F)	HFE	(E)T(F)	D	ETF	21
(E)(T)	(T)(F)	H(F)(E)	(T)		(E)(T)F	22
(T)(S)		(S)(T)			(T)	23
(T)(E)	(P)(T)	(H)(T)	(H)T(F)	A	T(F)(E)	24
V(E)TF	(P)FHT	V(E)TF	(H)T(F)	I	(E)TF	25
T(F)(E)		(T)(F)(E)	H(F)(T)	R	(T)(F)	26
(E)(T)	(F)	(E)VF	(F)H(E)		(E)TF	27
F(E)		HF(E)	(F)(E)		(E)TF	28
F(E)VT		H(F)(E)(V)		R	(E)TF	29
(V)F(E)	T(F)H(E)	(H)(F)(E)	EVFHT	E	(E)TF	30
(F)(E)	H(P)T	TF(E)	(F)H	A	(E)TF	31
T(F)		T(H)(F)	(F)(T)	C	TF	32
TF(E)	(P)T(E)	THF(E)	T(F)(E)	T	(E)TF	33
HF(E)T	(P)	HF(E)	(H)(T)(F)	I	(E)TF(V)	34
(T)(F)(E)	*(T)(F)*	*(T)(F)(E)*	*(E)(T)(F)*	V	*(E)T(F)*	35
T(F)(E)	*(F)T*	*T(F)(E)*	*(E)T(F)*	E	*(E)T(F)*	36
T(F)(E)	*T(F)*	*T(F)(E)*	*T(F)(E)*		*(E)T(F)*	37
T(F)(E)	*T(F)(E)*	*T*	*(E)FT(V)*		*(E)TF*	38
T(F)(E)	*(T)*	*T(F)*	*(T)(F)*	C	*T(F)(V)*	39
(T)(F)(E)	(T)(F)	(T)(F)H(E)	T(F)H(E)	O	(T)(F)(E)	40
				M		
(T)(F)(E)	(T)(F)(E)	THF(E)	(F)(T)(E)	P	(P)T(F)(E)	61
	(F)(E)(T)	VT(F)(E)	(E)TF(V)	O	(E)T(F)	62
(F)(E)(T)		(T)HF	(P)HF	U	P(E)T(F)	63
(F)(E)TV	(T)HF		(E)T(F)(V)	N	T(F)(E)	64
TF(E)(V)	(P)FH	(E)(V)T(F)		D	T(F)(E)	65
IN	AIR	———	———	S	———	66
(E)T(F)	(E)TP(F)	(E)T(F)	(E)T(F)			67

E = explosive (in varying degrees)
F = flammable/fire
G = gas emitted
H = heat given off
T = toxic gas emitted
P = polymerization
V = violent decomposition
S = solubilization of compounds

Note: This table predicts the events that will occur during the accidental or purposeful mixing of two or more chemicals. *Not all predictions are accurate or appropriate* due to interfering substances, additional materials in the mixture, variable concentrations of mixture components, and variable environmental conditions of the reactants. This table represents what *may happen* if the components are mixed appropriately under the right conditions.

CHEMICAL CLASS AND CHEMICAL REACTIVITY

CHEMICAL CLASS AND CHEMICAL REACTIVITY CHARACTERISTICS

1) A Chemical class separates the chemicals listed in alphabetical order in the previous section into reactivity groups according to molecular functional groups, chemical classes, or chemical reactivities. Each chemical group of specific chemical properties is assigned a "Reactivity Group Number" or RGN.

2) Groups with RGN 1 to 39 are based on molecular functional groups and reaction similarities within each group, whereas groups of chemicals with RGN 61 to 67 are based primarily on chemical reactivities of each group of chemicals.

3) All trade names in this section are denoted by asterisks (*) or trade marks (®) consistent with the notations used in the alphabetical listing of compounds.

4) This section is used to obtain the RGN of chemicals when the chemical constituents are known only by chemical classes, molecular functional groups, or chemical reactivities. A discussion of the toxicity level, physical and chemical characteristics, and decomposition characteristics of each chemical group is provided before each group of chemicals. Also, human sensitivity levels and significant levels of danger are usually discussed.

5) The listing was developed from the alphabetical listing of chemicals and from the references. Not all available chemicals are included in the listings. Additional chemicals will be added in the future.

6) If a specific chemical is missing from both of the lists, the group title (underlined) and group description can provide the information necessary for characterization in the "Reaction and Compatibility Chart." For example, a compound such as bromo-*sec*-butanol with *bromo-* (as a prefix) and *-ol* (at the end of the name) would suggest that the compound belongs in group (RGN) 17 and group (RGN) 4. The numbers (4 and 7) can be used to determine reaction possibilities with another number (or numbers) for a second compound in the "Chemical Reactivity and Compatibility Chart."

TABLE B

Reactivity Group Number	Group Name
1	Nonreactive Gases and Inert Gases
2	Inorganic Acids, Strong and/or Highly Toxic
3	Weak Acids including Organic Acids, Organic Anhydrides, Metal-acid Complexes, and Inorganic Acids
4	Alcohols, Glycols, Diols, Ethoxy-metal Compounds, and Stable Ethers
5	Aldehydes and Oximes
6	Amides, Imides, and Stable Urea Compounds
7	Amines, Anilines, Hydrazines, Pyridines, Triazines-Aliphatic, and Aromatic
8	Azo Compounds, Diazo Complexes, and Hydrazide Compounds
9	Carbamates and Thiourea Compounds
10	Caustics or Basic Compounds
11	Cyanides, Inorganic-type
12	Dithiocarbamates, Thiocarbamates, and Sulfur-containing compounds
13	Esters, Carboxamides, Oxalates, Carboxylates, Organo-borates, and Special Compounds
14	Ethers, Cyclic Ethers, Ethoxy-metal Compounds, and Special Reactives
15	Fluorides, Inorganic
16	Hydrocarbons, Aromatic
17	Halogenated Organics and Organochlorine Compounds
18	Isocyanates, Isothiocyanates, and Thiocyanates
19	Ketones
20	Mercaptans, Thiols, and Other Organic Sulfides
21	Metals, Nonmetals, Amalgams and Specific Alloys
22	Metals and Alloys—Dusts, Powders, Fume, and Vapors
23	Radioactive Poisons, Isotopes, Metals, Ions, Elements, and Compounds
24	Inorganic Compounds, Metals, Organometallic Substances, and Special Compounds

TABLE B (Continued)

Reactivity Group Number	Group Name
25	Nitrides
26	Nitriles, Cyanide-forming Organics, and Other Cyanide-containing Organics
27	Nitro Compounds
28	Hydrocarbon, Aliphatic, Unsaturated
29	Alkanes or Hydrocarbons, Aliphatic, Saturated
30	Organic Peroxides, Organic Hydroperoxides, and Organic Oxidizing Agents
31	Phenols, Creosols, Catechols, Resorcinol, Phenoxy-type Pesticides, and Stable Aromatic Ethers
32	Organophosphates, Phosphothiates, Phosphamides, and Phosphodithioates
33	Sulfides, Inorganic
34	Epoxides
35	Flammable Gases (Dangerous fire and explosion risk)
36	Flammable and Toxic Gases
37	Flammable, Corrosive Toxic, and Water-sensitive Gases
38	Oxidizing Gases
39	Toxic and/or Corrosive Gases, including Compressed or Liquified Gases
40	Human Carcinogenic Compounds and Dangerous Toxins (Chemical and Biological)
61	Combustible and Flammable Materials, Miscellaneous
62	Explosive Hazard and Shock-sensitive Compounds
63	Polymerizable Compounds
64	Oxidizing Agents, Oxidizers (Strong to Medium Strength) and Selective Oxidizers
65	Reducing Agents, Strong or Special Types of Reducing Agents
66	Air-reactive, Air-sensitive, and Self-reactive Substances
67	Water Reactive and Moisture-sensitive Compounds

RGN 1 NONREACTIVE AND INERT GASES

When heated, these gases tend to pressurize containers. Tanks of these gases can act as dangerous missiles when nozzles are broken off the pressurized tank or when a hole is punctured into the tank. Some of these gases can react in a fire to deplete oxygen and to form toxic substances. Generally, these gases can react with oxygen, fluorine, and chlorine at high temperature and pressure to form toxic gases. Nitrogen gas tends to be the most reactive with oxygen in a fire or in sunlight to form toxic gases. Some of these gases have radioactive isotope(s) that are used everyday or are a common problem.

Argon

Carbon dioxide

Helium (Helium-3)

Krypton (Krypton-86)

Neon

Nitrogen (Nitrogen-15)

Radon (Radon-222)

Rare gases

Xenon

RGN 2 INORGANIC ACIDS, STRONG AND/OR HIGHLY TOXIC

These acids are extremely corrosive, dangerous, and highly toxic. Acids react violently and explode in the presence of basic compounds and reducing agents. Avoid contact of acids with bases and reducing materials. A number of these acids are powerful oxidizing agents. Acids can react vigorously at ambient temperatures when stored in contact with organic and cellulose-type materials. Many acids will ignite combustibles (wood, paper, oil, cloth, and clothing, etc.) and other organic materials. Many of these acids can release toxic gases and/or flammable gases, such as NO_X, SO_3, and hydrogen. A number of these acids will evolve hydrogen on contact with most metals and are a dangerous fire risk. Each person must avoid breathing the vapors of these acids. Trained and experienced individuals should handle these chemicals properly.

Acids, inorganic, n.o.s.

Aqua Regia (Nitrohydrochloric acid)

Battery acid

Bromic acid

Caro's acid

Chloric acid

Chlorodifluoroacetic acid (Fairly strong)

Chlorosulfonic acid (Sulfuric chlorohydrin)

Chromosulfuric acid

Difluorophosphoric acid (Difluophosphoric acid)

Duclean*

Fluoroboric acid (Fluoboric acid)

Fluorophosphoric acid, anhydrous

Fluorosilicic acid (Fluosilic acid)

Fluorosulfonic acid (Fluosulfonic acid)

Hexafluorophosphoric acid

Hydrazoic acid (Hydrogen azide)

Hydriodic acid (Hydrogen Iodide; gas and aqueous solution)

Hydrobromic acid (Hydrogen Bromide; gas and aqueous solution)

Hydrochloric acid (Hydrogen Chloride; gas and aqueous solution)

Hydrofluoric acid (Hydrogen Fluoride; gas and aqueous solution)

Hydrogen azide (Hydrazoic acid)

Hydriodic acid (Hydrogen Iodide, gas and aqueous solution)

Iodic acid

Iodic acid anhydride (Iodine pentoxide)

Merchant acid

Mixed acid (Nitrating acid)

Monofluorophosphoric acid

Muriatic acid

Nitrating acid

Nitric acid

Nitrohydrochloric acid

Nitrosylsulfuric acid (Nitrosylsulphuric acid)

Nordhausen acid

Oleum (Oil of vitriol)

Perbromic acid

Perchloric acid

Perchlorous acid

Periodic acid

Permonosulfuric acid

Peroxysulfuric acid

Phosphoric acid

Phosphoric anhydride

Phosphorus acid (Phosphonic acid, ortho)

Phosphotungstic acid

Polyphosphoric acid

Pyrosulfuric acid

RFNA (Red fuming nitric acid)

Selenic acid

Strong Inorganic Acid Mists Containing Sulfuric Acid

Sulfuric acid (Oleum, Sulphuric acid)

Sulfuric acid, concentrated mists

Sulfurous acid (Sulphurous acid)

Tetraphosphoric acid

Trichloromethylphosphonic acid (Fairly strong)

Trifluoracetic acid (Fairly strong)

RGN 3 WEAK ACIDS INCLUDING ORGANIC ACIDS, ORGANIC ANHYDRIDES, METAL-ACID COMPLEXES, AND INORGANIC ACIDS.

Weak acids of various strengths are corrosive and flammable. Some organic acids are strong, but not usually as strong as the inorganic acids in RGN 2. A number of these acids are highly toxic to toxic. Additional acids are not extremely toxic and/or corrosive. Some of these acids can release flammable and/or toxic gases such as hydrogen, hydrogen cyanide, hydrogen sulfide, acid vapors, and other gases. A few of these acids are very poisonous and highly toxic, especially during decomposition and/or volatilization. Anhydrides added to water form acids. Each person must avoid breathing the vapors of these acids and anhydrides.

Acetamidoamino-2-naphthalenesulfonic acid

Acetic acid

Acetic anhydride

Acetoacetic acid

Acetylsalicylic acid (Aspirin, Aspirin dust)

Acid butyl phosphate

Acids, organic, n.o.s.

Acrylic acid (with or without sulfuric acid)

Adipic acid

Agent orange

Alkyl sulfonic acids (Alkyl sulphonic acids)

Alkyl sulfuric acids (Alkyl sulphuric acids)

Amino acid

4-Aminoazobenzene-3,4'-disulfonic acid (Acid yellow 9)

Aminobenzenedisulfonic acid

Aminobenzoic acid (Various forms)

Aminoethylsulfuric acid

Amino-G acid

Amino-J acid

4-Amino-1-phenol-2,6-disulfonic acid

para-Aminophenylmercaptoacetic acid

Aminopropionic acid

Aminosalicylic acid

Anisic acid

Anthranilic acid (ortho-Aminobenzoic acid)

Arsanilic acid (p-Aminophenylarsonic acid)

Arsenic acid (CCA*)

Aryl sulfonic acids (with and without sulfuric acid)

Aspirin dust

Azelaic acid

Bardische acid

Benzenephosphinic acid

Benzenephosphonic acid

Benzenesulphonic acid

Benzidinedicarboxylic acid

Benzoic acid

Blasticidin-S

Boric acid

Boron trifluoride acetic acid complex

Boron trifluoride propionic acid complex

Broenner acid
Bromoacetic acid
Bromobenzoic acid
Bromobutyric acid
Bromopropionic acid
Bromosuccinic acid
B.S. 500*
Butanoic acid
Butenoic acid
Butyl acid phosphate
Butyl peroxymaleic acid
Butyl peroxyphthalic acid
Butyric acid
C-acid
Cacodylic acid
Capric acid
Caproic acid
Caprylic acid
Carbamide phosphoric acid
CCA* (Arsenic acid)
CDTA
4-ChFu
Chicago acid (SS acid)
Chloramben
Chlorambucil
Chlorfenac
Chloroacetic acid
Chloroacetic anhydride
Chloroaminobenzoic acid (Various forms)
Chlorobenzoic acid

Chloroethylphosphonic acid
Chloromethylphenoxyacetic acid (MCPA)
Chloronitrobenzenesulfonic acid (Various forms)
Chloronitrobenzoic acid
Chlorophenoxyacetic acid
Chlorophenoxy herbicide(s)
Chlorophenoxypropionic acid
Chlorophthalic acid
Chloroplatinic acid
Chloropon
Chloropropionic acid (Various forms)
Chlorosalicylic acid
Chlorotoluenesulfonic acid (Various forms)
Chromic acid
Chromium trioxide
Chromotropic acid
Cinnamic acid
Cleve's acid
Cloprop
Clopyralid
CMPP
4-CPA
Crisalamina*
Crisamina*
Crocein acid
Crotonic acid
Cyanoacetic acid
Cyclamic acid
Cyclopentylpropionic acid
2,4-D

Dalapon

DAS

2,4-DB

2,4-DES

Diaminodiphenic acid

4,4'-Diamino-2,2'-stilbenedisulfonic acid (DAS)

Dianat*

1-Diazo-2-naphthol-4-sulfonic acid

Dibromomalonic acid

Dibromoterephthalic acid

Dicamba

Dicamix*

Dichloroacetic acid

Dichlorobenzoic acid

Dichlorophenoxyacetic acid (2,4-D)

Dichlorophenoxybutyric acid (2,4-DB)

2-(2,4-Dichlorophenoxy)propionic acid

Dichlorophthalic acid

Dichloroprop (Dichlorprop)

Dichloropropionic acid

Dichloroprop-P

Dichlorprop (Dichlorprop-P)

Diethylacetic acid

Di-(2-ethylhexyl)phosphoric acid

Diethylpyrocarbonate (Ester)

2-[4,5-Dihydro-4-methyl-4-(1-methylethyl)-5-oxo-1H-imidazol-2-yl]-
   3-quinolinecarboxylic acid

Dihydroxybenzoic acid

Diiodosalicylic acid

Diisooctyl acid phosphate

Dimethylarsinic acid
Dimethylolpropionic acid (DMPA)
Dinitrosalicylic acid
Diphenic acid
Diphenylacetic acid
Diphosphoric acid
Dithiocarbamic acid
Dodecylbenzenesulfonic acid
Dodine (Dodine acetate)
Doguadine
2,4-DP
DRC 1339
Erythritol anhydride (Butadiene dioxide)
Ethenoic acid
Ethephon (Ethrel*)
4-Ethylbenzenesulfonic acid
2-Ethylbutyric acid
Ethylenediaminetetraacetic acid (EDTA)
2-Ethylhexoic acid
Ethyl phosphoric acid
Ethylsulfuric acid (Ethylsulphuric acid)
Evisect*
Facet* (Fasnox)
F-acid
Fenac
Fenoprop
Fenton's reagent
Fluoroacetic acid
Fluorosulfamic acid
Formic acid

FP acids*
Fumaric acid
Furoic acid
Furylacrylic acid
G acid
Gallic acid
Garlon*
Gesapax-H
Glim*
Glutaric acid
Glycolic acid
Glyphosate-trimesium
G salt
H acid
Halazone
Heptafluorobutyric acid (Perfluorobutyric acid)
Heptanoic acid (Heptoic acid)
Hexahydrophthalic anhydride
Hexanedioic acid
Hexanoic acid
Hydrocinnamic acid
Hydrocyanic acid (Prussic acid; Liquid and Vapors)
Hydrofluorosilicic acid (Hydrofluosilicic acid)
Hydrogen cyanamide
Hydrogen cyanide (Gas or vapors from solution)
Hydrogen difluoride
Hydrogen selenide
Hydrogen sulfide
Hydrogen telluride
Hydroxyacetic acid

Hydroxybenzoic acid

Hydroxybutyric acid

Hydroxydibromobenzoic acid

Hydroxymethylbenzoic acid

Hydroxynaphthoic acid

2-Hydroxypropionic acid

8-Hydroxyquinoline benzoate

Hydroxysuccinic acid

Hypobromous acid

Hypochlorous acid

Hypophosphoric acid

Hypophosphorous acid

IAA (IA)

IBA

Imazaquin (Imazaquine)

Imazaquin, ammonium salt

Imazethapyr

Indole-3-acetic acid

Indole-3-butyric acid

Iodipamide

Isobutyric acid

Isobutyric anhydride

Isopentanoic acid

Isophthalic acid

Isopropyl acid phosphate

J acid

K acid

Koch acid

Lactic acid

Laurent's acid

Laurent's *alpha*-acid
Lentemul*
Levulinic acid
Lithic acid (Uric acid)
M-acid
Maleic acid
Maleic anhydride
Malic acid (Hydroxysuccinic acid)
Malonic acid
Mandelic acid
MCA
MCPA
MCPB
MCPP
MDBA
Mechlorethamine hydrochloride
Mecopex*
Mecoprop
Mecoprop-P
Melphalan
Mercaptoacetic acid
Mercaptopropionic acid
Metanilic acid
Metaxon
Methacrylic acid
Methanesulfonic acid
Methanoic acid
Methoxyacetic acid
Methylbenzoic acid
Methylbutenoic acid

2-(2-Methyl-4-chlorophenoxy)propionic acid
Methylphosphoric acid
Methylphthalanilic acid
Methylsulfonic acid
2M-4Kh-M
Monochloroacetic acid
Napalm
Naphthaleneacetic acid
Naphthenic acid
Naphthoxyacetic acid
Naphthylphthalamic acid
Neville-Winter acid (NW acid)
Nitrilotriacetic acid
Nitrobenzenesulfonic acid (Nitrobenzenesulphonic acid)
ortho-Nitrophenylpropiolic acid
N-Nitrososarcosine
Nitrous acid
Optima*
Orthoarsenic acid
Orthoboric acid
Oxalic acid
Oxygenated Hydrocarbon(s)*
PABA
Pamex*
PBA
Pelargonic acid
Penicillin
Peracetic acid
Performic acid
Peroxyacetic acid

Phenolsulfonic acid

Phenoxyacetic acid derivative pesticide

Phenylacetic acid

Phenyl *alpha*-acid

Phenyl J acid

Phenylpropionic acid

Phthalic anhydride

Picloram

Pivalic acid

Polychlorobenzoic acid

1,3-Propane sultone

Propanoic acid

Propel*

Propionic acid

Propionic anhydride

Propi-Rhap*

Proprop

Prussic acid, gas or liquid

Pyromellitic acid (PMA)

Pyromellitic dianhydride (PMDA)

Pyruvic acid

Quinclorac

R-acid

RR acid (2R acid)

S-acid

2-S acid (SS acid)

Salicylic acid dust

Scepter*

Schaeffer acid

Selenous acid (Selenious acid)

Shoelkopf acid
Shotgun*
Silvex (Fenoprop or 2,4,5-TP)
SS acid
Succinic acid
Sulfamic acid (Sulphamic acid)
Sulfobenzoic acid
2,4,5-T
Tarurine
2,3,6-TBA
TCA
Tecloftalam
Terephthalic acid
Thioacetic acid
Thiobenzoic acid
Thioclam hydrogen oxalate
Thiodipropionic acid
Thioglycolic acid
Thiolactic acid
Thiomalic acid
Thiosalicylic acid
TIBA (Triiodobenzole acid)
Tobias acid
Toluenesulfonic acid (Toluenesulphonic acid)
Toluic acid
Tolylenediaminesulfonic acid
Touchdown*
2,4,5-TP
TPA
Tribromoacetic acid (Fairly strong)

Tricamba
Trichloroacetic acid (Fairly strong)
Trichlorobenzoic acid
Trichlorophenoxyacetic acid
(2,4,5-Trichlorophenoxy)propionic acid
Triclopyr
Trihydroxybenzoic acid
Triiodobenzoic acid
Trimellitic acid
Trimellitic anhydride (TMA)
Trimethylacetic acid (Pivalic acid)
Trinatox D*
Trinitrobenzoic acid
2,6,8-Trioxypurine
Tri-Scept*
Tris(hydroxymethyl)acetic acid
TSA
Tsitrex
Undecylenic acid
Uric acid
Valeric acid
Vengador*
Vulcosal*
White acid
Xenic acid
Zobar*

RGN 4 ALCOHOLS, GLYCOLS, DIOLS, ETHOXY-METAL COMPOUNDS AND STABLE ETHERS (ALL ISOMERS)

These compounds are flammable and an explosion hazard. These vapors can be ignited by friction, heat, sparks, or flame. Vapors easily flash back from source of ignition. Containers of compounds may explode when heated. Many compounds evolve hydrogen gas when in contact with metals under heat. The alcohol compounds that contain bromo-, chloro-, cyano, and/or amino-groups are significantly more toxic and dangerous chemicals during a fire. In most cases, avoid breathing vapors of the compound and decomposition products from a fire or from heated compounds. Many of the compounds are toxic and can be absorbed by skin. Some of the larger molecules of alcohols are less flammable, less explosive and more stable.

Acetaldehyde cyanohydrin

Acetol

Acetone cyanohydrin

Alcohol, n.o.s. (Ethanol, Isopropanol, Methanol, etc.)

Allyl alcohol

2-Amino-1-butanol

Aminoethanol

Aminoethoxyethanol (Diglycolamine*, DGA)

Aminomethylpropanol (AMP)

Aminopropanol (Various forms)

Aminopropyldiethanolamine

AMP-95

Amyl alcohol(s)

Amyl ether

Anisic alcohol

Antifreeze

Antok*

BAL

Benzyl alcohol

Bishydroxycoumarin

Bis(hydroxyethyl)butynediol ether

Brodifacoum

Bromadiolone

2-Bromo-2-nitropropane-1,3-diol

Bronopol

Broprodifacoum

Bufotenine

Busulfan (1,4-Butanediol dimethylsulfonate)

Butanediol

Butanol (Butyl alcohol)

Butenediol

Butoxyethanol

Butoxyethoxypropanol

Butoxypolypropylene gylcol

Butoxypropanol

Butyl alcohol

Butylated hydroxyanisole (BHA)

Butyl "Cellusolve"*

Butylene glycol (Various forms)

1,4-Butynediol
Cannabis
Chloral hydrate (Knockout drops)
Chloramphenicol
Chlorobutanol
Chloroethanol
Chlorofenethol
Chlorohydrin
Chloropropanol
Clethodim
Crotyl alcohol
Cutless*
Cyanuric acid
Cyclohexanol
Cyclopentanol
Cyproconazole
DCPC
DDDM
Decanol
Denatured alcohol
DES
Diacetone alcohol
Diacetone alcohol peroxide
Dibenzyl ether
Dibromopropanol
Dibutylaminoethanol
Dichlorohydrin
Dichloroisopropyl alcohol
Dichlorophene (DDDM)
Di(*para*-chlorophenyl)ethanol (DMC, DCPC)

Dichloropropanol
Diclobutrazol
Diethanolamine (DEA)
Diethylaminoethanol
Diethylene glycol
Diethylpropanediol
Diethylstilbestrol (DES)
Diglycol chlorohydrin
Dihydroxynaphthalene (Naphthoresorcinol)
Diisopropanolamine (DIPA)
Diisopropyl carbinol
Diisopropylethanolamine
Dimercaptopropanol (BAL)
2-Dimethylaminoethanol (Deanol)
Dimethylaminopropanol (Various forms)
Dimethylethanolamine
Dimethylhexanediol
Dimethylhexynol
Dimethylisopropanolamine
Dimethyloctynediol
Dimethylolpropionic acid (DMPA)
Dimethylolurea
Dimethylpentanol (Various forms)
Diniconazole
Dipentaerythritol
Dipropylene glycol monomethyl ether
Ephedrine
2,3-Epoxy-1-propanol
Estradiol
Estriol

Ethanol (Ethyl alcohol)
Ethanolamine (MEA)
Ethoxyethanol
Ethyl alcohol
2-Ethylbutanol
Ethylbutyl alcohol
Ethylbutyl propanediol
Ethylene bromohydrin
Ethylene chlorohydrin
Ethylene cyanohydrin
Ethylene glycol
Ethylene glycol monobenzyl ether
Ethylene glycol monobutyl ether
Ethylene glycol monoethyl ether
Ethylene glycol monomethyl ether
Ethylhexanediol
Ethyl thioethanol
Flurprimidol
Flutriafol
Furfuryl alcohol
Fusel oil
Glycerin mist
Glycerol
Glycerol *alpha*-monochlorohydrin
Glycidol
Glycol monomethyl ether (Glycol monoethyl ether)
Glyoxal
Gossyplure
Heptanol(s)
Hexalin*

Hexamethylene glycol
Hexanediol
Hexanetriol
Hexanol(s)
Hexenol
Hexylene glycol
Hexynol
Hydroxyacetone
Hydroxyisobutyronitrile
Hydroxyproline
Hydroxypropanone
Hydroxypropionitrile
Impact
IPAE
Isoamyl alcohol
Isobutanol (Isobutyl alcohol)
Isooctyl alcohol
Isopentyl alcohol
lsopropanol
lsopropanolamine (MIPA)
lsopropanolamine dodecylbenzenesulfonate
Isopropoxyethanol (IPE)
Isopropyl alcohol
Isopropylaminoethanol (IPAE)
Lycopodium
Mercaptoethanol
Methac*
Methallyl alcohol
Methanol
Methoxybutanol

2-Methoxyethanol

Methoxypropanol

Methyl alcohol

Methylallyl alcohol

Methylamyl alcohol

alpha-Methylbenzyl alcohol

Methylbutanol

Methyl butynol

Methylcyclohexanol(s)

Methylisobutyl carbinol

Methylpentanol (Various forms)

3-Methylpentyn-3-ol

Methylpropanol (Various forms)

MIBC

Monoethanolamine

Monoisopropanolamine

Myleran®*

Neoflex*

2-Nitro-2-methyl-1,3-propanediol

Nonanol

Octanol

Pentaerythritol (PE)

Pentanol(s)

Propadrine hydrochloride

Propane sultone

1,3-Propane sultone

Propanol

Propanolamine

Propargyl alcohol

2-Propenol

Propoxypropanol

Propranolol

Propyl alcohol

Propylene chlorohydrin (1-Chloro-2-propanol)

Propylene glycol

Propylene glycol monomethyl ether

Purified Plus*

Pyrocatechol

Pyruvic alcohol

Quinhydrone

SBA

Scilliroside

SDA

Select*

Shellac (Lac, Gum lac, Stick lac; Alcohol solution)

Solvenol*

Tannic acid

Terpene alcohol

Tetraethanolammonium hydroxide

Tetramethyl-1,3-cyclobutanediol

Thiodiglycol

Thioglycol

Tribromo-*tert*-butyl alcohol

2,3,6-Trichlorobenzyloxypropanol

Trichloroethanol

Triethanolamine (TEA)

Triethanolamine dodecylbenzenesulfonate

Trimethylcyclohexanol

Trimidal (Triminol*)

Tris(hydroxymethyl)acetic acid

Tris(hydroxymethyl)aminomethane (THAM)

Tris(hydroxymethyl)nitromethane

Tritac*

Undecanol

Uniconazole-P (Uniconazole)

Unite*

Vigil*

Vinyl-2-ethylhexyl ether

Warfarin

Whiskey

Wine

Zoocoumarin

RGN 5 ALDEHYDES AND OXIMES (ALL ISOMERS)

Aldehydes are flammable or combustible. The compounds are a fire and/or explosion hazard indoors, outdoors, and in confined spaces such as sewers, basements, rooms, etc. Vapors usually flash back from the source of ignition. Vapors from these compounds can be ignited by friction, heat, sparks, or flame. Containers of compounds may explode when heated. In most cases, avoid breathing vapors of the compound and decomposition products from a fire or from heated materials. Some compounds can be absorbed by skin and are severe irritants. A few compounds can be lachrymators.

Acetaldehyde

Acetaldehyde ammonia

Acetaldehyde oxime

Acetaldol

Acrolein

Aldehyde ammonia

Aldehydes, n.o.s.

Aldol

Anisaldehyde

Aqualin

Benzaldehyde

Benzaldehyde cyanohydrin

Bromal

Butanal

Butenal

Butyraldehyde

Butyraldoxime

Chloral (Trichloroacetaldehyde)

Chloroacetaldehyde

Chloroacrolein

Chlorobenzaldehyde

2-Chloroethanal

Cinnamic aldehyde (Cinnamaldehyde)

Crotonaldehyde

3-Cyclohexane-1-carboxaldehyde

Decanal

Dibromosalicylaldehyde

Dichloroacetaldehyde

Dichlorobenzaldehyde

Dimethylaminobenzaldehyde

Dimethylpentaldehyde

Ethylene glycol diformate (Glycol diformate)

Ethylhexaldehyde(s)

2-Ethyl-3-propylacrolein

Formaldehyde

Formalin (In alcohol)

Furaldehyde(s)

Furfural

Furfuraldehyde

Glutaraldehyde

Glycidaldehyde

Glyoxal

Heptaldehyde

Heptanal

Hexaldehyde

Hexanal

Hydroxyadipaldehyde

Hydroxybenzaldehyde (Various forms)

Isobutyl aldehyde

Isobutyraldehyde

Isopentaldehyde

Magnacide* H

Metaldehyde

Methacrylaldehyde (Methacrolein)

Methanal

Methylpentaldehyde

alpha-Methylvaleraldehyde

Nonanal

Octadecenyl aldehyde

Octanal

Octyl aldehyde(s)

Paraformaldehyde

Paraldehyde

Pentanal

Propanal

2-Propenal

Propionaldehyde

Propyl aldehyde

Salicylal

Salicylaldehyde

1,2,3,6-Tetrahydrobenzaldehyde

4-Thiapentanal

Tolualdehyde

Tribromoacetaldehyde

Trioxane (Aldehyde properties)

Undecanal

Urea formaldehyde

Valeraldehyde

RGN 6 AMIDES, IMIDES, AND STABLE UREA COMPOUNDS (ALL ISOMERS)

These compounds tend to be flammable and produce toxic gases during fires and other decomposition processes. Many of these compounds can produce ammonia and carbon monoxide under a reductive atmosphere during a fire. During an oxygen-rich fire, nitrogen oxides are normally emitted. Many of these compounds are poisonous and highly toxic by ingestion and/or inhalation. Some of the compounds can be absorbed through the skin. Avoid breathing vapors of the compounds and products of decomposition during a fire or another event.

Acetamide

Acetanilide

Acetoacetanilide

Acetoacet-*ortho*-chloranilide

Acetoacet-*para*-phenetidide

Acetoacet-*para*-toluidide

Acetophenetidin

Acetylacetanilide

p-Acetylaminophenol (APAP)

Acetyl carbromal

Acrylamide

Amides, n.o.s.

Aminoacetanilide

Andalin* (Flucycloxuron)

ANTU

Assert*

Azobenzene

Azodicarbonamide (Azobisisobutyronitrile)

Barbital

Barbituric acid

Bayluscid*

Benzamide (Benzoylamide)

Benzoyl-2,5-diethoxyaniline

Bromacil

N-Bromoacetamide (NBA)

Bromobenzoyl acetanilide

Brucine

N-tert-Butylacrylamide

Butyramide

Caffeine (Theine)

Carbathiin

Carbetamide

Carboxin

Cascade*

CDAA

CDEA

Chloramphenicol

Chlorfenidim

Chloroacetamide

Chloroacetanilide

3-Chloro-4-benzamido-6-methylaniline

Chloro-N,N-diallylacetamide (CDAA)

Chloro-N-isopropylacetanilide

Chlorosalicylanilide

Chloroxuron
Cyanamide
Cyanoacetamide
Cyano(methylmercury)guanidine
Cyclobarbital
Cycloheximide
Cymoxanil
Cyprofuram
Dagger*
Deet
Diallylbarbituric acid
Diamate* (Chorphthalim)
Dibromsalan
Dichlormid
Dicryl
Diethylacetamide
Diethyldiphenylurea (Carbamite, Stabilizer)
N,N-Diethyl-*meta*-toluamide
Diflubenzuron
Diflufenican
Difolatan*
2-[4,5-Dihydro-4-methyl-4-(1-methylethyl)-5-oxo-1H-imidazol-2-yl]
   -3-quinolinecarboxylic acid
Dimethenamid
N,N-Dimethyl acetamide (DMAC)
N,N-Dimethylformamide (DMF)
5,5-Dimethylhydantoin (DMH)
Dimethylolurea
Dinitolmide
3,5-Dinitro-*o*-toluamide
Diphenamid

Di-*ortho*-tolylcarbodiimide
Di-*ortho*-tolylthiourea
Diuron (Dichlorfenidim)
DMAC
DMF
N-Ethylacetanilide
Euparen* (Dichlofluanid)
Fenuron
Flazasulfuron
Flufenoxuron
Flumiclorac-pentyl ester
Fluometuron
Fluoroacetamide
Fluoroacetanilide
Fluthiamide
Folpet
Formamide
Frontier*
Hexythiazox
Hydrogen cyanamide
Hydroxynaphthoic anilide
Imazaquin
Imazaquin, ammonium salt
Iprodion
Isocil (Isoprocil)
N-Isopropylacrylamide (NIPAM)
Isproturon (Crip*, Zodiac* TX)
Kerb*
Lenacil
Linuron
Lysergic acid diethylamide

Maleic hydrazide
Malonamide
Metobromuron
Metolachlor
MNFA
Monolinuron
Monuron (CMU)
Naphthaleneacetamide
Naphthylphthalamic acid
Naphthylurea
Neburon (Neburea)
Nialamide
Nomolt*
Norbormide
Noruron (Norea)
N-Octylbicycloheptene dicarboximide
Penicillin
Phenacaine hydrochloride
Phenacetin
Phenobarbital
N-Phenylacetamide
Phenylbarbital
Phenyl urea pesticide
Phthalimide derivative pesticide
Pronamide
Propachlor
Propalux*
Propanac*
Propanil*
Propanex*
Propionamide nitrile

Propylthiouracil (*RAHC*)

Propyzamide (Kerb*)

Prothidathion

Radiant*

Ramrod*

Randox*

Raticate

Resource*

Salicylanilide

Scepter*

Spike*

Strychnine

Strychnine salts

Sulfometuron methyl

Swep

Tebuthiuron

Teflubenzuron

Terbacil

Thioacetamide

Thiourea

3,4,5-Tribromosalicylanilide

Triforine

1,3,5-Triglycidyl-s-triazinetrione

Trimeturon

Tryparsamide

Valeramide

Zoalene

Zoamix*

RGN 7 AMINES, ANILINES, HYDRAZINES, PYRIDINES, TRIAZINES-ALIPHATIC AND AROMATIC

These compounds are normally reactive and corrosive. Most of these compounds are flammable, however some are combustible. These compounds can create fire and explosion hazards. Vapors can travel to ignition sources and flash back. Some of these compounds can be easily ignited by heat, sparks, or flame. Fire will produce toxic, irritating, and corrosive gases. These compounds are bases and usually combine with strong acids to form less dangerous salts. In an oxygen-depleted fire, these compounds can produce toxic substances. In addition, toxic forms of nitrogen oxides can be emitted when these compounds are in the presence of excess oxygen during a fire. Avoid skin contact and breathing the vapors of these substances and the products of decomposition. Most compounds are toxic and can cause severe injury or death. In some cases, dusts (solids) are toxic by inhalation. A number of compounds cause cancer in animals. Some compounds are highly explosive when combined with specific compounds in RGN 27 and/or RGN 62. Trained and experienced individuals should handle these chemicals properly.

Acetaldehyde ammonia

Acetaldehyde oxime

Acetamidoamino-2-naphthalenesulfonic acid

Acetoaminofluorene

ACN (ACNQ)

Acridine

Adamsite

Aerozine (UDMH)

Aldehyde ammonia

Alkaloids, n.o.s.

Alkylamines, n.o.s.

Allylamines, n.o.s.

Amdro*

Ametryn (Eviki*)

Amines, n.o.s. (Aliphatic and Aromatic)

Amino acid

Aminobenzene

Aminobenzoic acid

4-Aminobiphenyl

Aminobutane

2-Amino-1-butanol

Aminochlorophenol

Aminochlorotoluene

2-Amino-5-diethylaminopentane

p-Aminodiethylaniline

p-Aminodiethylaniline hydrochloride

p-Aminodimethylaniline

2-Amino-4,6-dinitrophenol

4-Aminodiphenyl

p-Aminodiphenylamine

Aminoethane

Aminoethanol

Aminoethoxyethanol

Aminoethylpiperazine

Aminomethane

Aminomethylpropanol (AMP)

Aminomethylpyridine

Aminomethylpyridine complexes
2-Aminonaphthalene
Aminopentane
Aminophenetole
Aminophenol
p-Aminophenylmercaptoacetic acid
2-Amino-3-picoline
Aminopropane
Aminopropanol
Aminopropionic acid
Aminopropionitrile
Aminopropyldiethanolamine
Aminopropylmorpholine
Aminopyridine(s)
Aminothiazole
Aminotoluene
3-Amino-1,2,4-triazole (Amitrole)
Amitraz
Amitrole
Amp-95
Amphetamine
Amylamine
Anilazine
Aniline
Aniline hydrochloride
Anisidine(s)
o-Anisidine hydrochloride
Anthranilic acid
Atrazine
Atropine

Avenge*
Avitrol*
Azidro*
Aziridine
Benzidine
Benzidinedicarboxylic acid
Benzidine dye
Benzidine sulfate
Benzocaine (Ethyl *para*-aminobenzoate)
Benzylamine
Benzyldimethylamine
Benzylpyridine
Bipyridilium pesticide, n.o.s.
Bis-(2-chloroethyl)ethylamine
Bis-(2-chloroethyl)methylamine
Bis(2,6-diethylphenyl)carbodiimide
Bis-(2-dimethylaminoethyl) ether (DMAEE)
2,4-Bis(isopropylamino)-6-methoxy-s-triazine
Blasticidin-S
Boron trifluoride monoethylamine
Brom 55*
Bromethalin
Brucine
Bufotenine
Butylamine
Butyl-*p*-aminobenzoate
Butylaniline
N,n-Butylimidazole
Cadaverine
Caprolactam, aerosol and vapor

Carbazole

Carbendazim

Carbenex*

Carzim*

CDTA

Chlorazine

Chlordiazepoxide hydrochloride

Chlordimeform (Galecron*)

Chloroaniline(s)

Chloroanisidine(s)

2-Chloro-4-ethylamino-6-isopropylamino-s-triazine

Chloromethylaniline

Chlorophenamidine

4-Chloro-o-phenylenediamine

Chloropyridine

Chlorotoluidine(s)

p-Chloro-o-toluidine

Chlorotoluidine hydrochloride

p-Chloro-o-toluidine hydrochloride

Chlorotrifluoromethylaniline

Clopidol

Coraza*

p-Cresidine

Cupferron

Cupriethylenediamine solution

Cyanopyridine

Cyclohexylamine

Cyprazine

Cyromazine

2,4-D, dimethylamine salt

Dapsone
DAS
Decylamine
Deiquat
DEN
Dialkylchloroalkylamine hydrochloride
Diallylamine
2,4-Diaminoanisole sulfate
Diaminoazoxytoluene
Diaminobenzene
Diaminobenzidine
Diaminobutane
Diaminodiphenic acid
Diaminodiphenylamine
Diaminodiphenyl ether
4,4'-Diaminodiphenylmethane (MDA)
3,3'-Diaminodipropylamine
Diaminohexane
Diaminophenol
Diaminopropane(s)
4,4'-Diamino-2,2'-stilbenedisulfonic acid (DAS)
2,4-Diaminotoluene
Di-*n*-amylamine
Diamylaniline
Dianisidine
Dianisidine diisocyanate
Dibenz[*a,h*]acridine
Dibenz[*a,j*]acridine
7*H*-Dibenzo[*c,g*]carbazole
N,N-Dibenzylamine

N,*N*-Dibenzylaniline
Dibutoxyaniline
Dibutylamine
Dibutylamine pyrophosphate
Dibutylaminoethanol
Di-*n*-butylaniline
Di-*sec*-butyl-*para*-phenylenediamine
Dicamba, dimethylamine salt
Dichloroaniline(s)
Dichlorobenzidine
3,3'-Dichlorobenzidine
3,3'-Dichlorobenzidine dihydrochloride
4,6-Dichloro-*N*-(2-chlorophenyl)-1,3,5-triazin-2-amine
3,3'-Dichloro-4,4'-diaminodiphenylmethane
3,5-Dichloro-2,4,6-trifluoropyridine
Diclobutrazol
Dicyclohexylamine
1,2-Di-(dimethylamino)ethane
Diethanolamine (DEA)
Diethoxyaniline
Diethylamine
Diethylaminoethanol
Diethylaminophenol
Diethylaminopropylamine
N,*N*-Diethylaniline
Diethylenediamine
Diethylenetriamine
N,*N*-Diethylethylenediamine
Di-(2-ethylhexyl)amine
Diethylnitrosoamine

p-Diethylnitrosoaniline

Diethyltoluidine

Difenzoquat methyl sulfate

Di-*n*-hexylamine

Diisobutylamine

Diisopropanolamine (DIPA)

Diisopropylamine

Diisopropylethanolamine

Dimenhydrinate

Dimepax*

Dimethametryn

Dimethoxyaniline

3,3'-Dimethoxybenzidine

Dimethoxychloroaniline

Dimethylamine, anhydrous (DMA)

Dimethylamine, aqueous soln.

Dimethylamine salt

Dimethylaminoacetonitrile

Dimethylaminoazobenzene (4-Dimethylaminoazobenzene)

Dimethylaminobenzaldehyde

Dimethylaminobenzenediazo sodium sulfonate (Dexon*)

2-Dimethylaminoethanol (Deanol)

2-Dimethylaminoethyl acrylate

Dimethylaminoethyl methacrylate

Dimethylaminopropanol

Dimethylaminopropylamine

N,N-Dimethylaniline

3,3'-Dimethylbenzidine

1,3-Dimethylbutylamine

Dimethylcyclohexylamine

Dimethylethanolamine

Dimethylhydrazine

1,1-Dimethylhydrazine (UDMH)

Dimethylisopropanolamine

Dimethylmorpholine

Dimethylnitrosoamine

Dimethylnitrosoaniline

Dimethyl-N-propylamine

Dimethylpyridine

Dinitramine

Dinitroaminophenol

Dinitroaniline

Diphenhydramine hydrochloride

Diphenylamine (DPA)

Diphenylamine chloroarsine

Diphenylbenzidine

Diphenylguanidine (DPG)

Dipicrylamine

Dipropylamine

Diquat (Diquat dibromide)

N,N'-Disalicylidene-1,2-diaminopropane

Di-ortho-tolylguanidine (DOTG)

Di-ortho-tolylthiourea (DOTT)

DM

DMA, anhydrous

DMN

Dodecylaniline

Dodine

Doguadine

Duacit*

Duomeen*

Dyrene*

Etaconazole

Ethanolamine

Ethylamine, gas or liquid

2-Ethylamino-4-isopropylamino-6-methylthio-s-triazine

Ethylaniline

Ethyl anthranilate

N-Ethyl-N-benzylaniline

N-Ethylbenzyltoluidine(s)

Ethylbutylamine (Various forms)

N-Ethylcyclohexylamine

Ethylenediamine

Ethyleneimine

2-Ethylhexylamine

N-2-Ethylhexylaniline

N-Ethylmorpholine

Ethylpicoline

Ethylpiperidine

N-Ethyl toluidine(s)

Ferimzone

Fipronil

Flufenzine

Flumetsulam

Fluoroaniline(s)

Furazolidone

Furfuryl amine

Gesapax-H

Guanidine

Guanidine carbonate

Guanidine hydrochloride

Guazatine

Herboxone*

Hexaconazole

Hexamethylenediamine

Hexamethyleneimine

Hexamethylenetetraamine (HMTA)

Hexamine

1,6-Hexanediamine

Hexanitrodiphenylamine (Hexite, Hexil)

Hexazinone

Hexetidine

Hexylamine

HN-1, (HN-2, HN-3)

Hydralazine hydrochloride

Hydramethylnon

Hydrazine, anhydrous

Hydrazine hydrate

Hydrazine nitrate

Hydrazine perchlorate

Hydrazines, n.o.s.

Hydrazine sulfate

Hydroxydiphenylamine

Hydroxyethylhydrazine

Hydroxylamine

Hydroxylamine hydrochloride

Hydroxylamine acid sulfate (HAS)

Hydroxylamine sulfate (HS)

N-beta-Hydroxypropyl-*ortho*-toluidine

Imibenconazole

Imidacloprid
3,3'-Iminodipropylamine
Imiprothrin
Impact
Indole
IPAE
Ipatone
Isobutylamine
Isophoronediamine
Isopropanolamine (MIPA)
Isopropanolamine dodecylbenzenesulfonate
Isopropylamine
Isopropylaminoethanol (IPAE)
Isopropylaniline
Isopropylpyridine
Isoquinoline
Jeffamine
London purple
2,6-Lutidine
Luxaquat*
Luxazim-F*
MBOCA (MOCA®)
MDA
MEA
Mechlorethamine hydrochloride
Mecoprop, dimethylamine salt
Melamine
Melphalan
Menazon
Menthanediamine

Mescaline

Metepa

Methenamine

Methoxyamine

Methoxyaniline

2-Methoxy-4,6-bis(isopropylamine)-s-triazine

Methoxypropylamine (3-MPA)

Methylamine, anhydrous

N-Methylaniline

5-Methyl-o-anisidine

Methylaziridine (2-Methylaziridine)

Methylbenzylamine

N-Methylbutylamine

Methyldiphenylamine

4,4'-Methylenebis(2-chloroaniline)

4,4'-Methylenebis(N,N-dimethylbenzenamine)

4,4'-Methylenedianiline

4,4'-Methylenedianiline dichloride

Methyl ethylpyridine (MEP)

Methylmorpholine (Various forms)

N-Methylpiperazine

1-Methylpiperidine

Methylpyridine(s)

Methylpyrrole

N-Methylpyrrolidine

N-Methyl-2-pyrrolidone

(Methylthio)aniline

Methyl yellow

Metribuzin

Mogeton G*

Monoethanolamine

Monoisopropanolamine

Monopropylamine(s)

Morpholine

3-MPA

MTD

MXDA

Myclobutanil

Naphthylamine

2-Naphthylamine

Nicotine (Nico Soap*)

Nicotine hydrochloride

Nicotine salicylate

Nicotine sulfate

Nicotine tartrate

Nitrapyrin

Nitroaniline

Nitrogen mustard

Nitrogen mustard hydrochloride

N-Nitrosodi-n-butylamine

N-Nitrosodiethanolamine

N-Nitrosodiethylamine

p-Nitrosodiethylaniline

N-Nitrosodimethylamine (DMNA, DMN)

p-Nitrosodimethylaniline

p-Nitrosodiphenylamine

N-Nitrosodi-n-propylamine

4-(N-Nitrosomethylamino)-1-(3-pyridyl)-1-butanone

N-Nitrosomethylvinylamine

N-Nitrosomorpholine

N-Nitrosonornicotine

N-Nitrosopiperidine

N-Nitrosopyrrolidine

N-Nitrososarcosine

Nitrotoluidine(s)

NNK

Nonylamine(s)

Norflurazon

Nornicotine

Octylamine

Organic pigment(s), self-heating

p,p'-Oxybis(benzenesulfonyl hydrazide)

4,4'-Oxydianiline

PABA

Paraquat (Herboxone*, Luxaquat*, Tenpar*)

Paraquat dichloride

Paraquat dihydride

1-Pentol

Pentylamine

Phenarsazine chloride

Phenazopyridine hydrochloride

Phenetidine(s)

Phenoxybenzamine hydrochloride

Phenylaniline

Phenylcarbylamine chloride

Phenylenediamine

2-Phenylethylamine

Phenyl J acid

N-Phenylmorpholine

N-Phenyl-alpha-naphthylamine

Phosgene oxime

Picoline(s)

4-Picolylamine
Picramic acid (Picraminic acid)
Picramide
Picridine
Pinacolyl methylphosphonofluoridate
Piperazine
Piperazine dihydrochloride
Piperidine
Pirimiphos-methyl (Pirimiphos-ethyl)
Polyalkylamines, n.o.s.
Polyamines, n.o.s.
Pramitol*
Primicid
Procarbazine hydrochloride
Prometon
Prometon 2,4-bis(isopropylamino)-6-methoxy-s-triazine
Prometryn
Propadrine hydrochloride
Propanediamine
Propanolamine
Propazine
Propiconazole
Propiodal
Propoxyphene
Propranolol
Propylamine
Propylenediamine
Propyleneimine
Pyranica (Tebufenpyrad)
Pyrazole
Pyrazoline

Pyrazoxyfen
Pyridine
Pyrifenox
Pyrrole
Pyrrolidine
Quinaldine
Quinine
Quinoclamine
Reglon
Saxitoxin
Semicarbazide hydrochloride
Shotgun*
Simazine (CAT)
Simeton
Simetryn
Sodium arsanilate
Spike*
Strychnine
Strychnine salts
Sulfometuron methyl
Sunter*
Tarurine
Tebuthiuron
TEM
Tenpar*
Tepa
Terbumeton
Terbuthylazine
Terbutryn (Ternit*)
Tetraconazole
Tetraethylenepentamine
Tetrahydrofurfurylamine

1,2,3,6-Tetrahydropyridine
Tetramethylenediamine
Tetramethylguanidine
Tetramine
Tetrazene
Thionazin
Thiosemicarbazide
TMA
o-Tolidine
2,4-Toluenediamine
Toluidine(s)
o-Toluidine
o-Toluidine hydrochloride
2,4-Toluylenediamine
Tolylenediaminesulfonic acid
Triallylamine
Triazine pesticide (Dyrene,* Anilazine)
Triazophos
Triazoxide
Tributylamine
Trichloromelamine
Triclopyr triethylamine salt
Tricyclazol
Trietazine
Triethanolamine (TEA)
Triethanolamine dodecylbenzenesulfonate
Triethylamine
Triethylenediamine
Triethylenemelamine (TEM)
Triethylenetetramine
Triflumizole

Trifluoromethylaniline

Triisopropanolamine

Trimethylamine, anhydrous

Trimethylamine, solution

Trimethylcyclohexylamine

Trimethylhexamethylenediamine(s)

Trinatox D*

Trinitroaniline

Tripelennamine citrate

Triphenylamine

Triphenylguanidine (TPG)

Tripropylamine

Tris(2-chloroethyl)amine

Tris(hydroxymethyl)aminomethane (THAM)

Tsitrex

TSPA

UDMH

Uniconazole-P (Uniconazole)

Vigil*

Vinylpyridine(s)

VX

Xanthine

m-Xylene-α, α'-diamine (MXDA)

Xylidine(s)

Yellow AB

Yellow OB

Zeatin

Zinc dimethyldithiocarbamate cyclohexylamine complex

Zinophos*

Zobar*

RGN 8 AZO COMPOUNDS, DIAZO COMPLEXES AND HYDRAZIDE COMPOUNDS

Most of these compounds are reactive, flammable, and somewhat corrosive. A number of compounds self-decompose or self-ignite when the compound is exposed to heat, friction, impact, or chemical reaction. Many of the self-reactive compounds are sensitive to heat and shock, while other compounds are acid sensitive. Some compounds decompose explosively and produce large amounts of gases when exposed to heat or fire or oxidizing materials. Avoid breathing the vapors of these substances and the products of decomposition. A number of compounds are toxic and/or cancer-causing agents. Experienced and properly trained individuals should handle these chemicals properly.

Aluminum tetraazidoborate

para-Aminoazobenzene (Aminoazobenzene)

para-Aminoazobenzene hydrochloride

Aminoazobenzenemonosulfonic acid

ortho-Aminoazotoluene

Azidocarbonyl guanidine

Azido-s-triazole

Azo compounds and Diazo compounds, n.o.s.

2,2'-Azodi-(2,4-dimethyl-4-methoxyvaleronitrile)

2,2'-Azodi-(2,4-dimethylvaleronitrile)

1,1'-Azodi-(hexahydrobenzonitrile

Azodiisobutyronitrile

2,2'-Azodi-(2-methylbutronitrile)

Benzenediazonium chloride

Benzene-1,3-disulfohydrazide (Benzene-1,3-disulphohydrazide)

Benzene-1,3-disulfohydrazine (Benzene-1,3-disulphohydrazine)

Benzene sulfohydrazide (Benzene sulphohydrazide)

Benzotriazole

4-[Benzyl(ethyl)amino]-3-ethoxybenzenediazonium zinc chloride

4-[Benzyl(methyl)amino]-3-ethoxybenzenediazonium zinc chloride

t-Butyl azidoformate

Chlorobenzotriazole

3-Chloro-4-diethylaminobenzenediazonium zinc chloride

DDN

DDNP

Diamine

Diazoaminobenzene

para-Diazobenzenesulfonic acid

Diazo compounds, n.o.s.

Diazodinitrophenol

2-Diazo-1-naphthol-4-sulfochloride (2-Diazo-1-naphthol-4-sulpho-
  chloride)

2-Diazo-1-naphthol-5-sulfochloride (2-Diazo-1-naphthol-5-sulpho-
  chloride)

1-Diazo-2-naphthol-4-sulfonic acid

4-Dimethylamino-6-(2-dimethylaminoethoxy)toluene-2-diazonium
  zinc chloride

Diphenyloxide-4,4'-disulfohydrazide (Diphenyloxide-4,4'-disulpho-
  hydrazide)

4-Dipropylaminobenzenediazonium zinc chloride

Guanyl nitrosoaminoguanylidine hydrazine

3-(2-Hydroxyethoxy)-4-pyrrolidin-1-yl benzenediazonium zinc chloride

Methyl hydrazine (MMH)

Phenylhydrazine

Phenylhydrazine hydrochloride

Sodium 2-diazo-1-naphthol-4-sulfonate (Sodium 2-diazo-1-naph-thol-4-sulphonate)

Sodium 2-diazo-1-naphthol-5-sulfonate (Sodium 2-diazo-1-naph-thol-5-sulphonate)

RGN 9 CARBAMATES AND THIOUREA COMPOUNDS

Many of these compounds are made up of groups containing one or more of the following substances: sulfur, chlorine, nitrogen, or bromine. During a fire, a number of these compounds can produce toxic forms of gases (hydrochloric acid, hydrobromic acid, sulfur oxides, phosgene, bromophosgene, hydrogen sulfide, ammonia, nitrogen oxides, and others) under the appropriate conditions of air supply. Avoid breathing the vapors from these compounds and the products of decomposition and/or reaction during a fire or another event. In several cases, use of a specific compound may be restricted. A few of the more complex carbamates are cholinesterase inhibitors.

Acetochlor

Alachlor

Alanycarb

Aldicarb

Aldoxycarb

Aminocarb

Azidro*

Barban

Bassa*

Baygon*

Bendiocarb
Benfuracarb
Benomyl
Betanal*AM
BPMC
Bufencarb
Butachlor
Butoxycarboxim
Carbamates, n.o.s,
Carbamate, pesticides n.o.s,
Carbamult
Carbaryl
Carbendazim (Carbendazime)
Carbenex*
Carbex*
Carbofuran (Furadan*)
Carbosulfan
Caribo*
Carzim*
Chloro-IPC (CIPC)
2-Chlorophenyl methylcarbamate
Chlorotoluron
6-Chloro-3,4-xylyl methylcarbamate
CPMC
Croneton*
Curasol*
Cypendazole
Decarbofuran
Desmedipham
2,6-Di-*tert*-butyl-*p*-tolyl-*N*-methylcarbamate

2,3-Dichloroallyl diisopropylthiolcarbamate

Dimepiperate

Dimethan

4-Dimethylamino-3-methylphenolmethyl carbamate

4-Dimethylamino-3,5-xylyl-N-methyl carbamate

3,4-Dimethylphenyl methylcarbamate

Dimetilan

Dioxacarb

Doubleplay*

Ethiofencarb

Ethyl carbamate

Ethylene thiourea

Fenobucarb

Formetanate (Formetanate hydrochloride)

Furadan*

IPBC

IPC (IPNC)

Isolan

Isoprocarb

Isopropoxyphenyl-N-methylcarbamate (Isopropoxyphenyl methyl-carbamate)

N-Isopropylmethylcarbamate

1-Isopropyl-3-methyl-5-pyrazolyldimethylcarbamate

Isopropyl N-phenylcarbamate

Kayabest*

Lannate* (Lanox*)

Luxazim-F*

Meprobamate

Mercaptodimethur

Mesurol*

Methasulfocarb

Methiocarb

Methomyl

2-(1-Methylethoxy)phenol methylcarbamate

4-(Methylthio)-3,5-dimethylphenyl-*n*-methylcarbamate

Metolcarb

Mexacarbate

MIPC

Mipcin*

Nac*

1-Naphthyl-*N*-methylcarbamate

Naphthylthiourea (ANTU)

Oncol*

Onic*

Oxamyl

Phenmedipham

Pirimicarb

Promecarb

Propamocarb hydrochloride

Propham (IFK)

Propoxur

2-*n*-Propyl-4-methylpyrimidyl-6-*N*,*N*-dimethylcarbamate

Pyrolan

Sanacarb*

Santochlor*

Sevimol*

Sevin*

Sta Brite* P

Standak*

Temik*

Thiocarboxime

Thiodicarb

Thiofanox (Thiofanocarb)

Trimethacarb

Urethane (Urethan; Ethyl carbamate)

Vanisect*

Vydate G*

XMC

Xylycarb

Xyly methylcarbamate

Yukamate*

RGN 10 CAUSTICS OR BASIC COMPOUNDS (STRONG BASE, INORGANICS AND ORGANIC AMINES)

These compounds are corrosive and reactive in water. They are acid-sensitive compounds, which are toxic. Avoid contact with acids. The ammonium compounds will evolve irritating fumes, especially when heated. Avoid breathing any vapors produced by these substances. Experienced individuals should handle these chemicals properly.

Alkali metal amides

Ammonia (Anhydrous ammonia)

Ammonium carbonate

Ammonium hydroxide

Anhydrous ammonia

Barium hydroxide

Barium oxide

Benzyltrimethylammonium methoxide

Beryllium hydroxide

n-Butylamine

Cadmium amide

Cadmium hydroxide

Caesium hydroxide

Calcium hydroxide

Caustics, n.o.s.

Caustic lime (Calcium hydroxide)

Caustic potash (Potassium hydroxide)

Caustic soda (Sodium hydroxide)

Cesium hydroxide

Chlorex (Sodium/potassium hypochlorite in basic solution)

Ethanolamine

Ethylamine

Guanidine carbonate

Lithium amide

Lithium hydroxide

Lithium methoxide

Lithium oxide

Lye (Sodium hydroxide, Potassium hydroxide)

Magnesium hydroxide

Methylamine

4-Picolylamine

Phenylethylamine

Potash

Potassium aluminate in water

Potassium amide

Potassium butoxide

Potassium hydroxide

Potassium hypochlorite in basic solution

Potassium oxide

Propylamine

Rubidium hydroxide

Soda lime

Sodium aluminate

Sodium amide (Sodamide)
Sodium carbonate
Sodium hypochlorite in basic solution
Sodium ethylate (Sodium ethoxide)
Sodium hydroxide
Sodium methoxide
Sodium methylate
Sodium oxide (Sodium monoxide)
Tetraethanolammonium hydroxide
Tetramethylammonium hydroxide
Tetramethylguanidine

RGN 11 CYANIDES, INORGANIC-TYPE

Cyanides form an extremely toxic gas in acids. Avoid contact with acids. Cyanides (poisons) are easily absorbed by skin. Solid cyanides must be handled with care. In aqueous solutions, cyanides should be mixed into solutions with a pH = 11 or above. Always avoid breathing any hydrogen cyanide vapors produced by cyanide compounds. Experienced individuals should handle these chemicals properly.

Barium cyanide

Barium cyanoplatinite

Black cyanide

Bromine cyanide (Cyanogen bromide)

Cadmium cyanide

Calcium cyanide

Carbonyl cyanide

Cobaltous cyanide

Cobalt potassium cyanide

Copper cyanide (Cupric cyanide)

Cuprous cyanide

Cuprous potassium cyanide

Cyanides, inorganic, n.o.s.

Cyanobrik*

Cyanogen (Gas or Liquid)
Cyanogen bromide
Cyanogen chloride (Gas or Liquid)
Cyanogen fluoride (Gas)
Cyanogen iodide
Cyanogran*
Gold cyanide
Hydrocyanic acid (Hydrogen cyanide, Prussic acid, Liquid)
Hydrogen cyanide, gas and vapors from solution
Lead cyanide
Mercuric cyanide (Mercury cyanide)
Mercuric oxycyanide
Mercuric oxycyanide, desensitized
Mercuric potassium cyanide
Mercury cyanide
Mercury oxycyanide, desensitized
Nickel cyanide
Potassium cuprocyanide (Potassium copper cyanide)
Potassium cyanide
Potassium gold cyanide
Prussic acid
Silver cyanide
Silver potassium cyanide
Sodium cuprocyanide (Sodium copper cyanide)
Sodium cyanide
Sodium gold cyanide
Zinc cyanide

RGN 12 DITHIOCARBAMATES, THIOCARBAMATES AND SULFUR-CONTAINING COMPOUNDS

These compounds are toxic and have varying degrees of flammability. A few compounds are noncombustible, while others are spontaneously combustible. Some compounds decompose violently when heated. The compounds contain sulfur and nitrogen groups which can form toxic gases (nitrogen oxide, ammonia, sulfur oxides, hydrogen sulfide, and others) during combustion under the appropriate conditions of a fire, explosion, accident, or similar incident. The dithiocarbamates have a tendency to form flammable gases (such as carbon disulfide) during fires. As a result, the carbon disulfide will probably produce an explosion risk during fires. Avoid breathing any vapors produced from these compounds or from the reaction and/or decomposition of these compounds.

Aminothiazole

Ammonium dimethyldithiocarbamate

Ammonium dithiocarbamate

Amobam

Avadex*

Benthiocarb

Butazate*

Butylate

Cadmium diethyldithiocarbamate

Carbamorph

Cartap
CDEC
Chloroallyl diethyldithiocarbamate
Copper based pesticide, n.o.s.
Copper dimethyldithiocarbamate
Cuprobam (Cuprobame)
Cycloate
Dazomet
Diallate (Di-allate)
Diammonium ethylenebisdithiocarbamate
N,N-Diethyldithiocarbamic acid 2-chloroallyl ester
Dimethyldithiocarbamic acid, zinc salts
Dimexano
Dinosulfon
Disodium ethylene-1,2-bisdithiocarbamate
Dithane*
Dithiocarbamates, pesticide, n.o.s.
Drepamon* (Tiocarbazil)
Duosan*
Enzone*
EPTC
Ethazcate*
Ethiolate
Ferbam
Fermate*
Flonex*MZ
IPX (Proxan)
Lanray*
Mancozeb
Mancozin*

Maneb (Manex*)

Manox* (Man-zox*)

Manzeb (Manzin*)

Mercury dimethyldithiocarbamate

Metacid* TS

Metam-sodium

Methazate*

Metiram

Milneb

Molinate

Nabam

Niacide*

Orbencarb

Padan

Parzate*

Pebulate

Phytex*

Polyram*

Potassium dimethyldithiocarbamate

Propineb

Prosulfocarb

Proxan (Proxane)

Raxil*T

Selenium diethyldithiocarbamate

Sodium dimethyldithiocarbamate

Sodium-N-methyldithiocarbamate

Sulfallate

Tepidone*

Thiadiazine

Thiobencarb

Thiocarbamate pesticide

Thiocarbamates, n.o.s.

Thiostat-B*

Thiostop*

Tiazon

Tillam*

Tiurante*

Tobosa*

S-2,3,3-Trichloroallyl-N,N-diisopropylthiolcarbamate

Trimanin*

Ultra-Clor*

Vegadex*

Vernolate (Vernam*)

Xanthate(s)

Xanthic acid

Yukamate*

Zerlate*

Ziman*

Zimate*

Zinc diethyldithiocarbamate

Zinc dimethyldithiocarbamate

Zinc dimethyldithiocarbamate cyclohexylamine complex

Zinc-1,2-propylene bisdithiocarbamate

Zinc salts of dimethyldithiocarbamic acid

Zineb*

Zipar*

Ziram*

RGN 13 ESTERS, CARBOXAMIDES, OXALATES, CARBOXYLATES, ORGANOBORATES AND SPECIAL COMPOUNDS

These compounds are normally stable and are either flammable or combustible. The compounds can be easily ignited by heat, sparks, or flames. Vapors form explosive mixtures with air. The vapors tend to travel back to the source of ignition and flash back. When heated, vapors form relatively easily and tend to explode in containers, sewers, basements, buildings, or outdoors. Fire can produce irritating, corrosive and toxic gases. Several compounds (especially the isobutyrates) and the decomposition products are toxic by ingestion, inhalation and/or skin absorption. In general, these compounds tend to be somewhat water-sensitive (reactive) and exhibit low to medium toxicity levels.

Acetoxybutane

Acetoxypentane

Acetylsalicylic acid

Acrylates, n.o.s.

Allethrin

Allyl acetate

Allyl chlorocarbonate

Allyl chloroformate

Allyl formate

Amyl acetate(s)

Amyl butyrate(s)

Amyl formate(s)

Amyl propionate

Areginal*

Assert*

Benzocaine

Bensoic derivative pesticide

Benzoin (Not benzoin resin)

Benzyl acetate

Benzyl benzoate

Benzyl chlorocarbonate

Benzyl chloroformate

Benzyl formate

Bifenthrin

Binapacryl

Bis(2-dimethylaminoethyl) ether (DMAEE)

Bis(2-ethylhexyl) phthalate

Brodifacoum (Bromadiolone)

Broprodifacoum

Butyl acetate

Butyl acrylate (Various forms)

Butylbenzyl phthalate

Butyl butyrate

Butyl chloroformate

tert-Butylcyclohexyl chloroformate

Butyl formate

Butyl lactate

Butyl methacrylate

Butyl phthalate

Butyl propionate(s)

Butrolactone

Candit*

2-Carbethoxycyclopentanone

Carbofos

Cellulose acetate

Cellulose compounds, fibers, pellets, granules, etc.

Chlorobenzilate*

Chloroethylchloroformate

Chloroformates, n.o.s. (All forms)

Chloromethyl chloroformate

Chlozolinate

Chrysanthemate compounds (Pyrethrins)

Cinerin I

Cinerin II

Cocaine

Croton oil

Cue-lure

Cyanoethyl acrylate

Cyanomethyl acetate

Cyclethrin

Cyclobutyl chloroformate

Cyclohexanol acetate

Cyclohexyl acetate

Cyclopropanecarboxylate

Dagger*

DCPA

DEHP

Diallyl maleate

Diallyl phthalate (DAP)

Dibutyl maleate (DBM)

Dibutyl oxalate

Dibutyl phenyl phosphate (DBPP)

Dibutyl phosphate

Dibutyl phthalate

Diclofop-methyl

Diethyl carbonate (Ethyl carbonate)

Diethyl diethylmalonate

Di(2-ethylhexyl) phthalate (DOP, DEHP)

Diethyl maleate

Diethyl phthalate

Difenacoum

Digital*

Diisodecyl phthalate (DIDP)

Dimethrin

Dimethyl carbate (Dimelone*)

Dimethyl carbonate

Dimethyl phthalate

Dimethyl terephthalate (DMT)

Dimethyl-2,3,5,6-tetrachloroterephthalate (DCPA)

Dinocap

Dinoseb acetate

Dinoterb acetate

Diphenyl phthalate

Diphosgene (Trichloromethyl chloroformate)

DOM

DOP

DPC

Erbon

Esters, n.o.s.

Ethofumesate

2-Ethoxyethyl acetate

Ethyl abietate

Ethyl acetate

Ethyl acetoacetate

Ethyl acrylate

Ethyl allylacetoacetate

Ethyl anthranilate

Ethyl borate

Ethyl bromoacetate

Ethylbromopyruvate

Ethyl butanoate

Ethylbutyl acetate

Ethylbutyl carbonate

2-Ethylbutyraldehyde

Ethyl butyrate

Ethyl chloroacetate

Ethyl chloroformate

Ethyl 2-chloropropionate

Ethyl chlorothioformate

Ethyl crotonate

Ethyl cyanoacetate

Ethylene glycol monoethyl ether acetate

Ethylene glycol monomethyl ether acetate

Ethyl formate

Ethyl formylpropionate

2-Ethylhexyl acrylate

2-Ethylhexyl chloroformate

Ethyl isobutyrate

Ethyl lactate

Ethyl methacrylate

Ethyl orthoacetate

Ethyl orthoformate

Ethyl orthopropionate

Ethyl oxalate

Ethylparaben

Ethyl propiolate

Ethyl propionate

Etoc*

Fenpropathrin

Fluazifop-*p*-butyl

Flucythrinate

Glycidyl acrylate

Glycol diacetate

Gossyplure

Haloxyfop-methyl

sec-Hexyl acetate

Hydroxyethyl methacrylate (HEMA)

Hydroxypropyl acrylate (HPA)

Isoamyl acetate

Isoamyl formate

Isobutyl acetate

Isobutyl acrylate

Isobutyl chloroformate

Isobutyl formate

Isobutyl isobutyrate (IBIB)

Isobutyl methacrylate

Isobutyl propionate

Isodecyl acrylate

Isooctyl thioglycolate

Isopropenyl acetate

Isopropyl acetate

Isopropyl butyrate

Isopropyl chloroacetate

Isopropyl chloroformate

Isopropyl 2-chloropropionate

Isopropyl formate

Isopropyl isobutyrate

Isopropyl propionate

Isoprothiolane

Kresoxim-methyl

Logico*

Luxathion*

Malathion (Malatop*)

Maldison

MCPA

MCPB

Medinoterb acetate

Mercaptothion (Mercaptotion)

Metalaxyl

Methac*

2-Methoxyethyl acetate

Methyl acetate

Methyl acetoacetate

Methyl acrylate

Methylamyl acetate

Methyl benzoate

Methyl bromoacetate

Methyl butyrate

Methyl carbonate

Methyl chloroacetate

Methyl chlorocarbonate

Methyl chloroformate

Methyl 2-chloropropionate

Methyl cyanoacetate

Methyl-2-cyanoacrylate

Methyl cyanoformate

Methyl formate

Methyl isovalerate

Methyl methacrylate

Methylparaben

Methyl pentanoate

Methyl propionate

Methyl salicylate

Methylvalerate

Oxygenated Hydrocarbon(s)*

Penncapthrin*

Pentyl acetate

Permethrin

Phenethyl propionate

Phenothrin

(3-Phenoxyphenyl)methyl(±)-cis,trans-3- (2,2-dichloroethenyl)-2,2-
   dimethylcyclopropanecarboxylate

2-Phenylethyl propionate

Phenyl salicylate

Piperalin

Pipron*

Prallethrin

Propargite

β-Propiolactone

Propoxyphene

Propyl acetate

Propyl butyrate

Propyl chloroformate

Propyl formate(s)

Pynamin*

Pyrellin*

Pyrethrins I (Pyrethrines)

Pyrethrins II (Pyrethrines)

Pyrethroid pesticides

Pyrethrum

Quizalofop-ethyl or Quizalofop-P-ethyl

Reserpine

Resmethrin

Sarin

Serinal

Smite*

Sodium chloroacetate

Sodium fluoroacetate

Stroby WG*

Superpalite (Green cross gas)

Tetralate*

Tetramethrin

Trichloromethyl chloroformate

2,4,5-Trichlorophenyl acetate

Triethyl borate

Triethyl phosphite

Triisopropyl borate

Trimedlure

Trimethyl borate

Trimethyl phosphite
Verdict*
Vinyl acetate
Vinyl butyrate
Vinyl propionate
Wintergreen oil

RGN 14 ETHERS, CYCLIC ETHERS, ETHOXY-METAL COMPOUNDS AND SPECIAL REACTIVES

Ethers are generally highly flammable and have a low flash point. Low molecular weight ethers (dimethyl ether, ethyl methyl ether, and diethyl ether or ethyl ether) are *extremely* flammable and are a severe fire and explosion hazard when exposed to heat and flame. Many ethers and ether-type compounds are unstable and/or are temperature sensitive. The compounds are easily ignited by heat, sparks, and flames. Vapors easily form explosive mixtures in air and in confined spaces. Vapors of many ether-type compounds travel to the source of ignition and flash back. Containers of the liquids may explode when heated. Avoid skin absorption and breathing any of these compounds. Overexposure can be fatal. Trained and experienced individuals should handle these chemicals properly. **Note: Low molecular weight ethers (i.e., dimethyl ether, methyl ether, methyl ethyl ether, etc.) contaminated with peroxides are extremely explosive.**

Acetal

Allyl ethyl ether

Allyl glycidyl ether

Anisole

Benzyl ether

Benzyl ethyl ether

Bis(chloromethyl) ether

Boron trifluoride diethyl etherate

Boron trifluoride dimethyl etherate
Bromodimethoxyaniline
Bromoethyl ethyl ether
Butoxyl
Butyl "Cellusolve"*
Butylene oxide
Butyl ether
Butyl ether, anhydrous with peroxides
Butyl magnesium chloride (In ether)
Butyl methyl ether
Butyl peroxydicarbonate
Butyl vinyl ether
Cetyl vinyl ether
Chloroethyl vinyl ether
Chloromethyl ether
Chloromethyl ethyl ether
Chloromethyl methyl ether
CMME
Chlomethoxyfen
DCIP
Diallyl ether
Dibutoxymethane
Dibutyl ether
Dichlorodiethyl ether
2,2-Dichloro-1,1-difluoroethyl methyl ether
Dichlorodiisopropyl ether (DCIP)
Dichlorodimethyl ether
Dichloroethyl ether
Dichloroisopropyl ether
Dichloromethyl ether

Diethoxymethane
3,3-Diethoxypropene
Diethylene dioxide
Diethyl ether
2,3-Dihydropyran
Diisopropyl ether
Dimethoxymethane
Dimethylacetal
Dimethyldioxane(s)
Dimethyl ether
Dimethylfuran
1,4-Dioxane (With or without peroxides)
Dioxolane
Diphenyl ether
Diphenyl oxide
Dipropyl ether
Divinyl ether
Ethers, n.o.s.
Ethoxyethanol
Ethyl butyl ether
Ethyl ether (Diethyl ether)
Ethyl ether containing peroxides
Ethylfuran
Ethyl methyl ether, gas or liquid
Ethyl propyl ether
Ethyl vinyl ether
Furan
Furfuran
Glycol ether
Glycol monoethyl ether

Grignard reagents in ether or tetrahydrofuran solvents
Halfenprox
Hexachloromethyl ether
Isopropyl ether
Isopropyl glycidyl ether (IGE)
Isopropyl peroxydicarbonate (Diisopropyl peroxydicarbonate)
IVE (Isobutyl vinyl ether)
Lithium aluminum hydride, etheral
Methylal (Formal)
Methyl *t*-butyl ether (MTBE)
Methyl chloromethyl ether
Methyl ether
Methyl ethyl ether
Methylfuran
Methyl propyl ether
Methyl vinyl ether (MVE)
MTBE
MTI-732
Nemamort*
Perfluoroethyl vinyl ether
Perfluoromethyl vinyl ether
Phenylether
Phenyl glycidyl ether (PGE)
Phenyllithium in benzene/ether
Polyglycol ether
Propyl ether
Propylpiperonyl ether
Tetrachloropropyl ether
Tetrahydrofuran (THF)
2,5-Tetrahydrofurandimethanol

Tetrol

THF

Thiophene

Trichloromethyl ether

Triisocyanatoisocyanurate of isophoronediisocyanate

Trinitroanisole

Trinitrophenyl methyl ether

Vinyl n-butyl ether

Vinyl 2-chloroethyl ether

Vinyl ether (Divinyl ether)

Vinyl ethyl ether (EVE)

Vinyl isobutyl ether (IVE)

Vinyl isopropyl ether

Vinyl methyl ether (MVE)

RGN 15 FLUORIDES, INORGANIC

Fluorides are corrosive and toxic. Many fluorides are moderately toxic and a strong irritant to tissue. Excess exposure to inorganic fluorides can irritate bones and cause fluorosis. Avoid breathing the vapors from any gaseous fluoride compound or the products of decomposition.

Aluminum fluoride

Ammonium bifluoride

Ammonium fluoborate

Ammonium fluoride

Ammonium fluorosilicate (Ammonium fluosilicate)

Ammonium hydrogendifluoride

Ammonium hydrogen fluoride

Ammonium perfluorooctanoate

Ammonium silicofluoride

Antimony fluoride

Antimony pentafluoride

Antimony trifluoride

Arsenic fluoride

Arsenic trifluoride

Arsenic pentafluoride

Barium fluoride

Beryllium fluoride

Bifluorides, inorganic, n.o.s.

Bismuth pentafluoride

Cadmium fluoride

Calcium fluoride

Calcium silicofluoride

Cerous fluoride

Cesium fluoride

Chromic fluoride

Chromium trifluoride

Cobaltous fluoride

Cobalt trifluoride

Copper fluoride (Cupric fluoride)

Ferric fluoride

Ferrous fluoride

Fluorides, inorganic, n.o.s.

Fluoroboric acid (Fluoboric acid)

Fluorophosphoric acid

Fluorosilicates, n.o.s.

Fluorosilicic acid

Fluorosulfonic acid (Fluosulphonic acid, Fluorosulfuric acid)

Fluosilicic acid

Gadolinium fluoride

Germanium potassium fluoride

Hexafluorophosphoric acid

Hydrofluoric acid

Hydrofluorosilicic acid (Hydrofluosilicic acid)

Hydrogen difluoride

Hydrogen fluoride, gas or liquid

Lanthanum fluoride

Lead fluoborate
Lead fluoride
Lithium fluoride
Magnesium boron fluoride
Magnesium fluoride
Magnesium fluosilicate (Magnesium fluorosilicate)
Magnesium silicofluoride
Manganic fluoride
Manganous fluoride
Mercuric fluoride
Molybendum hexafluoride
Potassium aluminum fluoride
Potassium bifluoride (Potassium acid fluoride)
Potassium fluoride
Potassium fluoroacetate
Potassium fluorosilicate
Potassium hydrogendifluoride
Potassium hydrogen fluoride
Potassium silicofluoride
Rubidium fluoride
Selenium fluoride
Selenium hexafluoride
Selenium tetrafluoride
Silicofluorides, n.o.s.
Silicon tetrafluoride, gas
Silver fluoride
Sodium acid fluoride
Sodium aluminum silicofluoride
Sodium bifluoride
Sodium fluoride

Sodium fluoroacetate
Sodium fluorosilicate
Sodium hydrogendifluoride
Sodium hydrogen fluoride
Sodium pentafluorostannate
Sodium silicofluoride
Stannous fluoride
Strontium fluoride
Sulfur hexafluoride (Sulphur hexafluoride), gas
Sulfur pentafluoride (Sulphur pentafluoride), gas
Sulfur tetrafluoride (Sulphur tetrafluoride), gas
Tantalum potassium fluoride
Tellurium hexafluoride
Thorium fluoride
Titanium potassium fluoride
Tungsten hexafluoride
Uranium tetrafluoride (Green salt)
Uranium hexafluoride
White acid
Ytterbium fluoride
Zinc fluoride
Zinc fluorarsenate
Zinc fluoroborate
Zinc fluorosilicate
Zinc silicofluoride
Zirconium ammonium fluoride
Zirconium potassium fluoride
Zirconium tetrafluoride

RGN 16 HYDROCARBONS, AROMATIC (ALL ISOMERS)

Hydrocarbons are highly flammable, while others can be combustible. Most hydrocarbons are easily ignited by heat, sparks, or flame. Hydrocarbon vapors form explosive mixtures in air. The vapors may travel to the source of ignition and flash back. Containers may explode when heated. The chemicals present a vapor explosion hazard in confined spaces and outdoors. Some hydrocarbons produce irritating, corrosive and/or toxic gases when burned.

Acenaphthene

Amylbenzene

Anthracene

Anthracene oil

Benz(e)acephenanthrylene

Benz[a]anthracene

Benzene

Benzo[b]fluoranthene, Benzo[j]fluoranthene and Benzo[k]fluoranthene

Benzo[a]pyrene (Benzopyrene)

Benzo[e]pyrene

Benzylbenzene

p-Benzylphenol

Biphenyl

n-Butylbenzene

Butyltoluene(s)

Chlorotoluene(s)

Chrysene

Cumene

Cyclo Sol*

Cymene(s) (Cymol)

Decyl benzene

Dibenz[a,h]anthracene

Dibenzo[a,e]pyrene

Dibenzo[a,h]pyrene

Dibenzo[a,i]pyrene

Dibenzo[a,l]pyrene

Diethylbenzene

Diisopropylnaphthalene

Diphenyl

Diphenylacetylene

Diphenylethane

Diphenylethylene

Diphenylmethane

Dodecylbenzene

Ethylbenzene

Hemimellitene

Hexamethylbenzene

Hydrocarbons, aromatic, n.o.s.

Indene

Indeno[1,2,3—c,d—]pyrene

Isobutylbenzene

Isodurene

Isopropenylbenzene
Isopropylbenzene
Isopropylnaphthalene
Isopropyltoluene
Mesitylene
Methylbenzene
5-Methylchrysene
Methyl naphthalene
Methyl styrene
Musk xylene
Naphthalene
Naphthaleneacetamide
Naphthaleneacetic acid
Naphthol
Nitroanisole
o-Nitroanisole
Nitrobenzene
Nitrobiphenyl
6-Nitrochrysene
Nitronaphthalene
Nitrotoluene(s)
Nitroxylene(s)
Oil mist (Polynuclear aromatic hydrocarbons, PAH)
PAHs
Particulate Polycyclic aromatic hydrocarbons (PPAH)
Pentamethylbenzene
Phenanthrene
Phenylbenzene
Phenylbutane
Phenylethane

Phenyllithium in benzene/ether
Phenylpropane
Polycyclic aromatic hydrocarbons
Polystyrene
Propylbenzene
Pseudocumene
Safrole
Stilbene
Styrene
Terphenyl
Tetramethylbenzene
Tetraphenylethylene
Toluene
Toluol
1,2,4-Trimethylbenzene
1,3,5-Trimethylbenzene
Trimethylbenzene (Various forms)
Triphenyl ethylene
Triphenylmethane
Vinylbenzene
Vinyl toluene
Xylene

RGN 17 HALOGENATED ORGANICS AND ORGANOCHLORINE COMPOUNDS (ALL ISOMERS)

Many halogenated organics are highly flammable and may be ignited by heat, sparks, or flame. Lower molecular weight organic-halide compounds are a dangerous fire hazard and tend to exhibit significant explosion limits in air. Some halogenated organics are combustible. Halogenated organics are toxic or highly toxic to fatal. Vapors have a tendency to travel to the source of ignition and flash back. The vapors can be an explosion hazard in sewers, confined spaces, and outdoors. Containers can explode when heated. At times, a few compounds are hydrolyzed by moisture producing acid vapor emissions. The compounds tend to produce acid gases (hydrochloric acid, hydrobromic acid, hydrofluoric acid, and others) during fires. The products of fire will be toxic, corrosive, irritating, and can cause severe burns or injury to the body. Some of the compounds exhibit carcinogenic properties. A few compounds, such as ethylene chlorohydrin, can penetrate ordinary rubber gloves and protective clothing. Avoid skin absorption and breathing vapors from these compounds and their combustion by-products.

Acetamiprid

Acetoacet-*ortho*-chloranilide

Acetochlor

Acetyl bromide

Acetyl carbromal

Acetyl chloride

Acetylene tetrabromide

Acetyl iodide

ACN (ACNQ)

Agent orange (Contains traces of chlorinated dioxins)

Alachlor

Aldrin

Allyl bromide

Allyl chloride

Allyl chlorocarbonate

Allyl chloroformate

Allyl iodide

Allyltrichlorosilane

Aluminum alkyl halides, n.o.s.

Amdro*

Aminochlorophenol

Aminochlorotoluene

Ammonium hexachloroplatinate

Amyl chloride

Amyltrichlorosilane

Andalin* (Flucycloxuron)

Anilazine

Anisoyl chloride

Aramite*

Aroclor* or Arochlor* (Aroclor® 1254, Aroclor® 1260)

Askarel

Atrazine

Avadex*

Axall*

Azidithion

Barban

Bayclean*
Bayluscid*
Baythroid*
BCNU
Benefin
Benthiocarb
Benz(e)acephenanthrylene
Benzal bromide
Benzal chloride
Benzalkonium chloride
Benzene hexachloride
Benzethonium chloride
Benzoepin
Benzo[b]fluoranthene
Benzo[j]fluoranthene
Benzo[k]fluoranthene
Benzotribromide
Benzotrichloride
Benzotrifluoride
Benzoyl chloride
Benzoyl fluoride
Benzyl bromide
Benzyl chloride
Benzyl chlorocarbonate
Benzyl chloroformate
Benzyl chlorophenol
Benzyl dichloride
Benzyl fluoride
Benzylidene chloride
Benzyl iodide

Beret*

BFE

BHC (Benzene hexachloride)

Bifenthrin

2,2-Bis(p-bromophenyl)-1,1,1-trichloroethane

Bis(2-chloroethyl)ethylamine

Bis(2-chloroethyl)methylamine

Bis(2-chloroethyl)nitrosourea

Bis(2-chloroethyl)sulfide

Bis(2-chloroethyl)sulphide

Bis(chloromethyl)ether

Bis(p-chloromethoxy)methane

2,2-Bis(p-chlorophenyl)-1,1-dichloroethane (TDE)

1,1'-Bis(p-chlorophenyl)-2,2,2-trichloroethanol (Keltane)

Bis(tetrachloroethyl)disulfide

Bis(trichlorosilyl)ethane

Bithionol

Boron bromodiiodide

Boron dibromoiodide

Boron tribromide

Boron trichloride

Boron trifluoride

Boron trifluoride, dihydrate

Boron trifluoride acetic acid complex

Boron trifluoride diethyl etherate

Boron trifluoride dimethyl etherate

Boron trifluoride monoethylamine

Boron trifluoride propionic acid complex

Boron triiodide

Brittox*

Brodifacoum
Bromacil
Bromadiolone
Bromal
Bromchlophos
Bromethalin
Bromine azide
Bromine chloride
Bromine cyanide
Bromine monofluoride
Bromine pentafluoride
Bromine trifluoride
Bromlost
N-Bromoacetamide (NBA)
Bromoacetic acid
Bromoacetone
Bromoacetone cyanohydrin
Bromoacetyl bromide
Bromoacetylene
Bromobenzene
Bromobenzoic acid
Bromobenzyl cyanide
Bromobenzyl trifluoride
Bromobutane
Bromobutyric acid
Bromochlorodifluoromethane
3-Bromo-1-chloro-5,5-dimethylhydantoin
Bromochloroethane
Bromochloroethene
Bromochloromethane

Bromochloropropane
Bromochlorotrifluoroethane
Bromodichloromethane
Bromodiethylaluminum
Bromodimethoxyaniline
Bromoethyl chlorosulfonate
Bromoethyl ethyl ether
Bromofluorobenzene
Bromoform
Bromol
Bromomethane
1-Bromo-3-methylbutane
Bromomethylethyl ketone
Bromomethylpropane(s)
2-Bromo-2-nitropropane-1,3-diol
2-Bromopentane
Bromophenol
Bromophenylphenol
Bromophos
Bromophosgene
Bromopicrin
2-Bromopropane (all bromopropanes)
Bromopropene (all bromopropenes)
Bromopropionic acid
Bromopropyne (all bromopropynes)
Bromosilane
Bromosuccinic acid
N-Bromosuccinimide, dry (NBS)
Bromothiophenol
Bromotoluene

Bromotrichloromethane

Bromotrifluoroether

Bromotrifluoroethylene, gas or liquid

Bromotrifluoromethane (Bromotrifluomethane)

Bromoxynil

Bronopol

Broprodifacoum

Brotal*

Bulan

Bulldock*

Butachlor

Butonate (Butichlorofos)

Butyl bromide

Butyl chloral hydrate

Butyl chloride

Butyl chloroformate

tert-Butylcyclohexyl chloroformate

Butyldichloroarsine

Butyl dichloroborane

Butyl fluoride

Butyltin trichoride

Butyltrichlorosilane

Butyryl chloride (Butyroyl chloride)

Camphechlor

Captab

Captafol*

Captan (Captane)

Carbon tetrabromide

Carbon tetrachloride

Carbon tetrafluoride

Carbon tetraiodide

Carbonyl chloride

Carbonyl fluoride

Carbonyl sulfide (Carbonyl sulphide)

Carbophenothion

Cascade*

Castellan*

CBM

CCNU

CDAA

CDEA

CDEC

CECA

Celest*

4-ChFu

Chinmix*

Chlomethoxyfen

Chloracetyl chloride

Chloral (Trichloroacetaldehyde)

Chloral hydrate

Chloralose

Chloramben

Chlorambucil

Chloramine

Chloramine-T

Chloramizo

Chloramphenicol

Chloranil

Chlorazine

Chlorbenside

Chlordane

Chlordecon (Kepone*)

Chlordiazepoxide hydrochloride

Chlordimeform

Clorendic acid

Chlorethoxyfos

Chlorfenac

Chlorfenapyr

Chlorfenidim

Chlorfenvinphos

Chlorinated camphene (Toxaphene)

Chlorinated dioxins

o-Chlorinated diphenyl oxide

Chlorinated naphthalene (Two chlorines and higher)

Chlorinated paraffins (C_{12}, 60% Chlorine)

Chlorine pentafluoride

Chlorine trifluoride

Chlormephos

Chlormequat chloride

Chloroacetaldehyde

Chloroacetamide

Chloroacetanilide

Chloroacetic acid

Chloroacetic anhydride

Chloroacetone

Chloroacetonitrile

Chloroacetophenone

Chloroacetyl chloride

Chloroacetylene

Chloroacrolein

Chloroacrylonitrile

Chloroallyl diethyldithiocarbamate (CDEC)

Chloroaminobenzoic acid

Chloro-*tert*-amylphenol

Chloroaniline(s)

Chloroanisidine(s)

Chlorobenzal

Chlorobenzaldehyde

3-Chloro-4-benzamido-6-methylaniline

Chlorobenzene

Chlorobenzilate*

Chlorobenzoic acid

Chlorobenzotriazole

ortho-Chlorobenzotrichloride

Chlorobenzotrifluoride(s)

Chlorobenzoyl chloride

Chlorobenzene

Chlorobenzilate

1-(2-Chloroethyl)-3-cyclohexyl-1-nitrosourea

Chlorfenvinphos

Chlorobenzene

Chlorobenzotriazole

Chlorobenzotrifluorides

Chlorobenzoyl peroxide

Chlorobenzyl chloride(s)

Chlorobenzyl cyanide

Chlorobenzylidene malononitrile (CS, OCBM)

Chlorobromomethane

1-Chloro-3-bromopropane (Chlorobromopropanes)

Chlorobutadiene

Chlorobutane

Chlorobutanol

Chlorobutyronitrile

Chlorocresol(s)

Chlorodecone

Chloro-*N*,*N*-diallylacetamide (CDAA)

Chlorodiborane

3-Chloro-4-diethylaminobenzenediazonium zinc chloride

Chlorodifluoroacetic acid

Chlorodifluorobromomethane

Chlorodifluoroethane(s)

Chlorodifluoromethane

Chlorodiisobutyl aluminum

Chlorodimethylamine diborane

Chlorodinitrobenzene

Chlorodinitrotoluene

Chlorodiphenyl

Chlorodipropyl borane

1-Chloro-2,3-epoxypropane

2-Chloroethanal

Chloroethane

Chloroethanol

Chloroethoxyfos

2-Chloro-4-ethylamino-6-isopropylamino-s-triazine

Chloroethylchloroformate

Chloroethylene

Chloroethylenimine

1-(2-Chloroethyl)-3-(4-methylcyclohexyl)-1-nitrosourea (MeCCNU)

Chloroethylphosphonic acid

Chloroethyl vinyl ether

Chlorofenethol

Chlorofenvinphos

Chlorofluorocarbon

Chloroform

Chloroformates, n.o.s.

Chloroformoxime

Chlorofos*

Chloroheptane(s)

Chlorohexane(s) (Dichlorohexane, Trichlorohexane, etc.)

Chlorohydrin

Chloro-IPC

Chloro-*N*-isopropylacetanilide

Chloromethane

Chloromethylaniline

Chloromethylchloroformate

Chloromethylchlorosulfonate

Chloromethyl ether

Chloromethyl ethyl ether

Chloromethyl methyl ether

Chloromethylnaphthalene

Chloromethylphenol

Chloromethylphenoxyacetic acid (MCPA)

4- Chloro-*o*-phenylenediamine

3-Chloro-4-methylphenyl isocyanate

Chloromethylphosphonic dichloride

3-Chloromethylpropene

Chloronaphthalene oil

Chloroneb

Chloro-*m*-nitroacetophenone

Chloronitroaniline(s)

Chloronitrobenzene(s) (*ortho-*, *meta-*, *para-*)

Chloronitrobenzenesulfonic acid

Chloronitrobenzoic acid

Chloronitrobenzotrifluoride

Chloronitropropane

Chloronitrotoluene(s)

Chloropentafluoroacetone

Chloropentafluoroethane

Chloropentane

Chloroperoxybenzoic acid

Chlorophacinone

Chlorophenamidine

Chlorophenate(s) (e.g., Potassium chlorophenate)

Chlorophenolate(s)

Chlorophenol(s)

Chlorophenoxyacetic acid

Chlorophenoxy herbicide (2,4-D; 2,4-DB; Dichlorprop; Erbon*; Falone*; MCPA; MCPB; Mecoprop; Silvex; 2,4,5-T)

Chlorophenoxypropionic acid

Chlorophenyl isocyanate(s)

2-Chlorophenyl methylcarbamate

3-(*p*-Chlorophenyl)-5-methylrhodanine

Chlorophenylphenol

Chlorophenyl phenylsulfone (Chlorodiphenyl sulfone)

Chlorophenyltrichlorosilane

Chlorophos

Chlorophthalic acid

Chloropicrin

Chloropivaloyl chloride

Chloroplatinic acid

Chloropon

Chloroprene (Gas or Liquid)

Chloropropane

Chloropropanol

Chloropropene (Various forms, Gas or Liquid)

Chloropropionic acid

Chloropropionitrile

Chloropropylene oxide

Chloropropyl mercaptan

Chloropropyne

Chloropyridine

Chlorosalicylanilide

Chlorosalicylic acid

Chlorosilane(s)

Chlorostyrene

N-Chlorosuccinimide (NCS)

Chlorosulfonic acid (Chlorosulphonic acid)

Chlorotetrafluoroethane

Chlorothalonil

Chlorothene*

Chlorothiophenol

Chlorothymol

Chlorotoluene(s)

Chlorotoluenesulfonic acid

Chlorotoluidine(s)

p-Chloro-o-toluidine

Chlorotoluidine hydrochloride

p-Chloro-o-toluidine hydrochloride

Chlorotoluron

2-Chloro-6-(trichloromethyl)pyridine

Chlorotrifluoroethane

Chlorotrifluoroethylene

Chlorotrifluoromethane

Chlorotrifluoromethylaniline

Chlorotrinitrobenzene

Chlorovinyldichloroarsine

Chlorovinylmethylchloroarsine

para-Chloro-*meta*-xylenol

Chloroxuron

Chloroxynil

Chloroxylenol

6-Chloro-3,4-xylyl methylcarbamate

Chlorpicrin

Chlorpromazine

Chlorpyrifos (Chlorpyriphos)

Chlorquinox

Chlorthion

Chlorthiophos

Chlozolinate

Cinnamoyl chloride

Clopidol

Cloprop

Clopyralid

CMME

CMPP

Cobra*

Concep* III

Copper chlorotetrazole

Copper trifluoroacetylacetonate

Coroxon

Coumaphos
Coumarin
Coumarin derivative pesticide
4-CPA
CPMC
Crisalamina*
Crisamina*
Croneton*
Crotyl bromide
Crotyl chloride
Crufomate
CS gas (Aerosol)
Curalin*
Cutless*
Cyanazine
Cyanochloropentane
Cyanogen bromide
Cyanogen chloride (Gas or Liquid)
Cyanogen fluoride (Gas)
Cyanogen iodide
(S)-Cyano(3-phenoxyphenyl)methyl-(S)-chloro-*alpha*-(1-methylethyl)benzeneacetate
Cyanuric chloride
Cyclobutyl chloroformate
Cyclohexenyltrichlorosilane
Cyclohexyl chloride
Cyclohexyl trichlorosilane
Cyclopentyl bromide
Cyclopentylpropionyl chloride
Cyclophosphamide

Cuclosal* (Cycloprothrin)

Cyfluthrin

Cyhalotrin

Cyhexatin

Cypermethrin (Cypermethrine)

Cyprazine

Cyproconazole

Cyprofuram

2,4-D (Chlorophenoxy compounds)

2,4-D, dimethylamine salt

Dalapon

2,4-DB

DBC

DBCP

DCB

DCIP

DCNA

DCPA

DCPC

DDD

DDDM

DDE

DDH

DDT

DDVP

2,4-DEB

Decabromobiphenyl

Decamethrin

Deiquat

Delsam*

Deltamethrin

2,4-DEP

2,4-DES

DFDD

DFDT

Dialifor

Dialkylchloroalkylamine hydrochloride

Diallate (Di-allate)

Diamate* (Chlorphthalim)

Dianat

Diazepam

Dibenzyldichlorosilane

Dibromoacetylene

Dibromobenzene(s)

1,2-Dibromobutan-3-one

Dibromochloromethane

Dibromochloropropane(s) (DBCP)

1,2-Dibromo-3-chloropropane

Dibromodiethyl sulfide

Dibromodifluoromethane

1,3-Dibromo-5,5-dimethylhydantoin

Dibromoethane

1,2-Dibromoethane

Dibromofluorobenzene

Dibromoformoxime

3,5-Dibromo-4-hydroxybenzonitrile

Dibromomalonic acid

Dibromomethane

Dibromomethyl ether

Dibromopropanol

Dibromosalicylaldehyde

Dibromoterephthalic acid

Dibromotetrafluoroethane

Dibromsalan

Dicamba

Dicamba, dimethylamine salt

Dicamix*

Dichlobenil

Dichlofenthion

Dichlone

Dichloramine-T

Dichlormid

Dichloroacetaldehyde

Dichloroacetic acid

Dichloroacetone

Dichloroacetyl chloride

Dichloroacetylene

2,3-Dichloroallyl diisopropylthiolcarbamate

Dichloroamine

Dichloroaminophenol

Dichloroaniline(s)

Dichloroanisole

Dichlorobenzaldehyde (Various forms)

Dichlorobenzalkonium chloride

Dichlorobenzene(s)

1,4-Dichlorobenzene

Dichlorobenzidine

3,3'-Dichlorobenzidine

3,3'-Dichlorobenzidine dihydrochloride

Dichlorobenzoic acid(s)

Dichlorobenzonitrile

Dichlorobenzotrichloride

Dichlorobenzoyl chloride

2,4-Dichlorobenzoyl peroxide

Dichlorobenzyl chloride

1,1-Dichloro-2,2-bis(*para*-chlorophenyl)ethane (TDE)

1,1-Dichloro-2,2-bis(*para*-ethylphenyl)ethane (Perthane*)

Dichlorobutane (DCB, Various forms)

Dichlorobutene (Various forms)

4,6-Dichloro-*N*-(2-chlorophenyl)-1,3,5-triazin-2-amine

Dichloro-(2-chlorovinyl)arsine

3,3'-Dichloro-4,4'-diaminodiphenylmethane

Dichloro-5,6-dicyanobenzoquinone (DDQ)

Dichlorodiethyl ether

Dichlorodiethyl sulfide (Mustard gas)

Dichlorodifluoroethane

Dichlorodifluoroethylene (Dichlorodifluoroethene)

2,2-Dichloro-1,1-difluoroethyl methyl ether

Dichlorodifluoromethane (Fluorocarbon-12)

Dichlorodiisopropyl ether (DCIP)

Dichlorodimethyl ether

1,3-Dichloro-5,5-dimethylhydantoin

Dichlorodimethylsilane

Dichlorodiphenyldichloroethane (TDE)

Dichlorodiphenyldichloroethylene (DDE)

Dichlorodiphenyltrichloroethane (DDT)

Dichloroethane (Various forms)

1,2-Dichloroethane

Dichloroethene

Dichloroethyl acetate

Dichloroethylarsine

Dichloroethylene

Dichloroethyl ether

Dichloroethylformal

1,1-Dichloro-1-fluoroethane

Dichlorofluoromethane (Fluorocarbon-21)

Dichloroformoxime

Dichlorohydrin

Dichloroisocyanuric acid

Dichloroisocyanuric acid salts

Dichloroisopropyl alcohol

Dichloroisopropyl ether

Dichloromethane

Dichloromethylchloroformate

Dichloromethyl ether

Dichloromethylsilane

Dichloromonofluoromethane

Dichloronitrobenzene (Various forms)

1,1-Dichloro-1-nitroethane

Dichloropentane(s)

Dichlorophenarsine hydrochloride

Dichlorophene (DDDM)

Dichlorophenol

Dichlorophenoxyacetic acid (2,4-D)

Dichlorophenoxybutyric acid (2,4-DB)

Di-(4-chlorophenoxy)methane

2-(2,4-Dichlorophenoxy)propionic acid

O-(2,4-Dichlorophenyl)-O,O-diethyl phosphorothioate

Di(*para*-chlorophenyl)ethanol (DMC, DCPC)

Dichlorophenyl isocyanate(s)

o-(2,4-Dichlorophenyl)-*o*-methyl isopropylphosphoramidothioate (DMPA)

2,4-Dichlorophenyl-4-nitrophenyl ether

Dichlorophenyltrichlorosilane

Dichlorophthalic acid

Dichloroprop (Dichlorprop)

Dichloropropane(s)

Dichloropropanol(s)

Dichloropropene

1,3-Dichloropropene (Technical grade)

Dichloropropionic acid

Dichloroprop-P

Dichloropropylene

Dichloropyrenes

Dichlorosilane

Dichloro-s-triazine-2,4,6-trione

Dichlorotetrafluoroacetone

1,2-Dichloro-1,1,2,2-tetrafluoroethane (Fluorocarbon-114)

Dichlorotetrafluoroethane

Dichlorotoluene

3,5-Dichloro-2,4,6-trifluoropyridine

Dichlorovinylchloroarsine

Dichlorovinyldimethyl phosphate

Dichlorovinylmethylarsine

Dichlorprop (Dichlorprop-P)

Dichlorvos (Dichlorovos, DDVP)

Diclobutrazol

Diclofop-methyl

Diclomezine

Dicloran

Dicofol

Dicryl

Dictran*

Dieldrin

Dienochlor

Diethyl chlorophosphate

Diethyl chlorovinyl phosphate

Diethyl-1-(2,4-dichlorophenyl)-2-chlorovinyl phosphate

Diethyl dichlorosilane

N,N-Diethyldithiocarbamic acid 2-chloroallyl ester

Diethylgermanium dichloride

O,O-Diethyl phosphorochloridothiate (Ethyl PCT)

O,O-Diethyl-O,3,5,6-trichloro-2-pyridylphosphorothioate

Difenoconazole

Diflubenzuron

Diflufenican

Difluorochloroethane(s)

Difluorodibromomethane

Difluoroethane

Difluoroethylene

Difluoromethane

Difluorophosphoric acid (Difluophosphoric acid)

Difluron

Difolatan*

Diglycol chlorohydrin

2,2'-Dihydroxy-5,5'-difluorodiphenyl sulfide

Diiodacetylene (Diiodethyne)

Diiododibromoethylene

Diiododiethyl sulfide

Diiodosalicylic acid

Dilan

Diman A

Dimefox

Dimenhydrinate

Dimension*

Dimethenamid

Dimethoxychloroaniline

Dimethylchloroacetal

Dimethyl chlorothiophosphate

Dimethyl-1,2-dibromo-2,2-dichloroethyl phosphate

Dimethyldichlorosilane

Dimethyldichlorovinyl phosphate

O,O-Dimethyl phosphorochloridothioate (Methyl PCT)

Dimethyl-2,3,5,6-tetrachloroterephthalate (DCPA)

Dimethyl thiophosphoryl chloride

Dimethyltin dichloride

Dimethylvinyl chloride

Diniconazole

Dinitramine

Dinitrobenzoyl chloride

Dinitrochlorobenzene (Dinitrochlorbenzene)

Dinitrofluorobenzene (DNFB)

Dioxin (Tetrachlorodibenzo-p-dioxin)

Diphenylamine chloroarsine (DM)

Diphenylbromoarsine

Diphenylchloroarsine

Diphenyldichlorsilane

Diphenylmethyl bromide

Diphosgene

Dipterex*

Diquat (Diquat dibromide)
Disulfuryl chloride
Dithiopyr
Diuron
DM
Dodecyltrichlorosilane
Doom*
Doubleplay*
2,4-DP
Drazoxolon
DRC 1339
Drinox*
Duacit*
Du Nema*
Dursban*
Dylox*
Dyrene*
Eleeter*
Endosulfan
Endrin
Enflurane
Enilconazole
Epibromohydrin
Epichlorohydrin
Epoxyheptachlor (HCE)
Erbon
Esfenvalerate
Etaconazole
Ethalfluralin
Ethanoyl chloride

Ethchlorvynol
Ethephon (Ethrel*)
Ethiofencarb
Ethrel*
Ethylbenzyl chloride
Ethyl bromide
Ethyl bromoacetate
Ethylbromopyruvate
Ethyl chloride, gas or liquid
Ethyl chloroacetal
Ethyl chloroacetate
Ethyl chloroformate
Ethyl 2-chloropropionate
Ethyl chlorosulfonate
Ethyl chlorothioformate
Ethyl dichloroarsine
Ethyl dichlorosilane
Ethylene bromohydrin
Ethylene chlorohydrin
Ethylene dibromide
Ethylene dichloride
Ethyl fluoride
Ethylhexyl chloride
2-Ethylhexyl chloroformate
Ethylidene chloride
Ethylidene fluoride
Ethyl iodide
Ethyl iodoacetate
Ethyl PCT
N-Ethyl perfluorooctane sulfonamide

Ethylphenyldichlorosilane
Ethyl phosphonothioc dichloride, anhydrous
Ethyl phosphonous dichloride, anhydrous
Ethyl phosphorodichloridate
Ethyltrichlorosilane
Etridiazole
Euparen* (Euparen*M)
Facet* (Fasnox*)
Falone*
Faneron*
Fenac
Fenazaflor
Fenbuconazole
Fenchlorphos (Fenchlorfos)
Fenophosphon
Fenoprop
Fenpiclonil
Fenson
Fentichlor
Fenvalerate
Fipronil
FireMaster BP-6
FireMaster FF-1
Flazasulfuron
Flocoumafen
Fluazifop-*p*-butyl
Fluazinam
Fludioxonil
Flufenoxuron
Flufenzine

Flumetralin
Flumetsulam
Flumiclorac-pentyl ester
Fluometron
Fluorbenside
Fluoroacetamide
Fluoroacetanilide
Fluoroacetic acid
Fluoroacetophenone
Fluoroaniline(s)
Fluorobenzene
Fluoroboric acid (Fluoboric acid)
Fluorodichloromethane
Fluorodifen
Fluoroform
Fluorogesarol
Fluoroimide
Fluoromethane
Fluorophenol
Fluorosulfamic acid
Fluorotoluene(s)
Fluorotrichloromethane
Fluquinconazole
Flurprimidol
Flusilazole
Fluthiamide
Flutriafol
Fluvalinate
Fluxofenim
Folpet

Fomesafen
Fopel
Fortress*
FP acids
Freon*
Frontier*
Fthalide
Full*
Fumaryl chloride
Furoyl chloride
Fury*
Gambit*
Garlon*
gamma-BHC
Genesolv*
Gesapax-H
Gix
Glim*
Glycerol alpha-monochlorohydrin
Halazone
Halfenprox
Halogenated organics, n.o.s.
Halon*, gas
Halothane
Haloxyfop-methyl
HCCH
HCE
HCH
HEOD
Heptachlor

Heptachlor epoxide

Heptafluorobutyric acid

Heptafluoropropane

Heptanoyl chloride

Heptenophos

Hercon* Vaportape II (Dichlorovos)

Hexabromobiphenyl

Hexabromoethane

Hexachloroacetone

Hexachlorobenzene

Hexachlorobutadiene

Hexachlorocyclohexane (Various forms; BHC, HCCH, HCH, TBH, Lindane, Benzene hexachloride)

Hexachlorocyclopentadiene

Hexachlorodiphenyl oxide

Hexachloroethane (Perchloroethane)

Hexachloromethylcarbonate (Triphosgene)

Hexachloromethyl ether

Hexachloronaphthalene

Hexachlorophene (Hexachlorophane)

Hexachloropropylene (Hexachloropropene)

Hexaconazole

Hexadecyltrichorosilane

Hexafluoroacetone

Hexafluoroacetone hydrate

Hexafluoroethane (Fluorocarbon 116)

Hexafluorophosphoric acid

Hexafluoropropylene (Perfluoropropene)

Hexafluoropropylene oxide (HFPO)

Hexaflurate

Hexanoyl chloride
Hexyltrichlorosilane
Hexythiazox
HFPO
HHDN
HN-1 (HN-2, HN-3)
Hostaquick*
Hydramethylnon (Amdro*, Siege*)
Hydroxydibromobenzoic acid
Hydroxymercurichlorophenol
Imazalil
Imibenconazole
Imidacloprid
Impact*
Iodine monobromide (Iodine bromide)
Iodine monochloride (Iodine chloride)
Iodine pentafluoride
Iodine trichloride
Iodipamide
Iodoacetic acid, sodium salt
Iodobutane
Iodoform
Iodomethylpropane
Iodopropane
N-Iodosuccinimide
Ioxynil
Iprodione
Isazofos
Isoamyldichloroarsine
Isobenzan

Isobutyl chloroformate

Isobutyryl chloride

Isocil (Isoprocil)

Isocyanobenzotrifluorides

Isodrin

Isophthaloyl chloride

Isopropenylchloroformate

Isopropyl bromide

Isopropyl chloride

Isopropyl chloroacetate

Isopropyl chloroformate

Isopropyl 2-chloropropionate

Isopropyl iodide

Isopropyl methylphosphonofluoridate

Isopropyl methylphosphoryl fluoride

Isotox*

Kanechlor®

Kelthane

Kepone®

Kerb*

Korax

L (Lewisite)

Lactofen

Lanray*

Lentemul*

Lewisite

Lindane

Linuron

Lithate* 2,4-D

Logico*

Lucel*
Luxafan*
Malachite Green
Marlate*
Martonite*
Maxim*
Maytril*
MBOCA
MCA
MCPA (Chlorophenoxy herbicide)
MCPA-thioethyl
MCPB
MCPCA
MCPP
MDBA
Mecopex*
Mecoprop
Mecoprop, dimethylamine salt
Mecoprop-P
Medallion*
Melphalan
Merbromin
2-Mesitylenesulfonyl chloride
Metaxon
Methoxychlor (DMDT)
Methylallyl chloride (MAC)
Methyl bromide
Methyl bromoacetate
Methyl bromoacetone
Methyl chloride

Methyl chloroacetate
Methyl chlorocarbonate
Methyl chloroform
Methyl chloroformate
Methyl chloromethyl ether
2-(2-Methyl-4-chlorophenoxy)propionic acid
Methyl 2-chloropropionate
Methylchlorosilane
Methylchlorosulfonate
Methyldichloroacetate
Methyldichloroarsine
Methyl dichlorosilane
4,4-Methylenebis(2-chloroaniline)
Methylene bromide
Methylene chloride
Methylene iodide
Methyl ethyl chloride
Methyl fluoride
Methyl fluorosulfonate
Methyl iodide
Methylmercury pentachlorophenate
Methyl phencapton
Methylphenyldichlorosilane
Methyl phosphonic dichloride
Methyl phosphonous dichloride
Methyl trichloroacetate
Methyltrichlorosilane
Methylvinyldichlorosilane
Metobromuron
Metolachlor

Mikado*

Mirex

2M-4Kh-M

MNFA

Mogeton G*

Monacide*

Monochloroacetic acid (Chloroacetic acid)

Monochloroacetone (Chloroacetone)

Monochlorobenzene

Monolinuron

Monuron (CMU)

Morfamquat (Morfamquat dichloride)

Mospilan*

MTI-732

Mustard gas

Myclobutanil

Mycotox

Naled

Neburon (Neburea)

Nemamort*

Nendrin

Neostomosan*

Nichlorfos

Nitrapyrin

Nitrobenzotrifluorides

Nitrobenzoyl chloride

Nitrobromobenzene

Nitrochlorobenzene

3-Nitro-4-chlorobenzotrifluoride

Nitrochloroform

Nitrofen

Nitrogen mustard (Various forms)

Nitrogen mustard hydrochloride

Nitrotrichloromethane

Nogos*

Nomolt*

Nonyltrichlorosilane

Norflurazon

Nuvan*

Octabromobiphenyl

Octachloronaphthalene

Octadecyltrichlorosilane

Octafluorobut-2-ene

Octafluorocyclobutane

Octafluoropropane

Octanoyl chloride

Octyl iodide

Octyl trichlorosilane

Orbencarb

Organochlorine pesticide, n.o.s.

Ornalin*

Oxadiazon

Oxalyl chloride

Oxsol 100*

Oxyfluorfen

Oxygen difluoride (Oxygen fluoride)

Padan

Pamex*

Paraquat dichloride

PBA

PBBs

PCBs

PCNB (Pentachloronitrobenzene)

PCP

PDB

Pelargonyl chloride

Penncapthrin*

Pentac*

Pentachloroethane (Pentalin)

Pentachloronaphthalene

Pentachloronitrobenzene (PCNB)

Pentachlorophenol (PCP)

Pentafluoroethane

Pentanochlor

Perchlorobenzene

Perchloroethane

Perchloroethylene (PERC)

Perchloromethyl mercaptan

Perfluidone

Perfluoroethyl vinyl ether

Perfluoroisobutylene

Perfluoromethyl vinyl ether

Permethrin

Perthane*

Phenacyl bromide

Phenacyl chloride

Phenacyl fluoride

Phenarsazine chloride

Phenothiol

(3-Phenoxyphenyl)methyl(±)-cis,trans-3-(2,2-dichloroethenyl)-2,2-
dimethylcyclopropanecarboxylate

Phenylacetyl chloride

Phenylcarbylamine chloride

Phenyl chloroformate

Phenylchloromethyl ketone

2-Phenyl-6-chlorophenol

Phenyldichloroarsine

Phenylhydrazine hydrochloride

Phenylphosphorus dichloride

Phenylphosphorus thiodichloride

Phenyltrichlorosilane

Phosalone

Phosgene

Phosgene oxime

Phosnichlor

Phosphamidon

Phosphoric bromide

Phosphoric chloride

Phosphorus oxybromide

Phosphorus oxychloride

Phosphorus pentabromide

Phosphorus pentachloride

Phosphorus pentafluoride

Phosphorus tribromide

Phosphorus trichloride

Phosphorus triiodide

Phosphoryl bromide

Phosphoryl chloride

Phthalide

Phthaloyl chloride

Picloram

Picryl chloride

Piperalin

Pipron*

Pirate*

Platinic sal ammoniac

Polirac*

Polybrominated biphenyl(s)

Polychlor

Polychlorcamphene

Polychlorinated biphenyl(s)

Polychlorinated dioxin(s) (PCDDs)

Polychlorinated triphenyl(s)

Polychlorobenzoic acid

Polyhalogenated biphenyl(s)

Polyhalogenated terphenyl(s)

Polytetrafluoroethylene (PTFE)

Polyvinyl chloride (PVC)

Polyvinyl fluoride

Potassium dichloroisocyanurate

Potassium dichloro-s-triazinetrione

Procure*

Profenofos

Prolan

Pronamide

Propachlor

Propalux*

Propamocarb hydrochloride

Propanac*

Propanil (Propanex*)

Propanoyl chloride

Propargyl bromide

Propargyl chloride

Propazine

Propiconazole

Propiodal

Propionyl chloride

Propi-Rhap*

Proprop

Propyl bromide

Propyl chloride

Propyl chloroformate

Propyl chlorosulfonate

Propylene chlorohydrin

Propylene dichloride

Propyltrichlorosilane

Propyzamide (Kerb*)

Prothiofos

Pyraclofos

Pyranica

Pyrazoxyfen

Pyridaben

Pyridate

Pyrifenox

Pyrinox*

Pyrosulfuryl chloride

Quinclorac

Quinoclamine

Quintozene

Quizalofop-ethyl (Quizalofop-P-ethyl)

Rabicide

Radiant*

Ramrod*

Randox*

Reglon

Resource*

Responsar*

Rizolex*

Ronilan*

Ronnel

SAGA*

Santochlor*

Saphire*

Sarin

Scout X-TRA

Septiphene

Serinal*

Sesone (SES)

Shotgun*

Silicochloroform

Silvex

Simazine

Smite*

Sneezing gas

Sodium chloroacetate

Sodium-2,4-dichlorophenoxyacetate

Sodium dichloro-s-triazinetrione

Sodium pentachlorophenate

Sodium pentachlorophenoxide

Sodium-2,3,4,6-tetrachlorophenate

Sodium trichloroacetate (Sodium TCA)

Soilfume*

Soman
Sonalan*
Storm* Rodenticide (Storm*)
Stratagem*
Sulcotrione
Sulfallate
Sulfonyl chloride
Sulfonyl fluoride
Sulfur bromide
Sulfur chloride
Sulfur dichloride
Sulfur monochloride
Sulfur oxychloride
Sulfuryl chloride
Sulfuryl fluoride, gas
Sulphenone
Sulphur chloride
Sulphuryl chloride
Sulphuryl fluoride, gas
Superpalite
Swep
2,4,5-T
2,4,6-T
2,3,6-TBA (TCBA)
TBH
TCA
TCB
TCDD (Tetrachlorodibenzo-*p*-dioxin)
TCNA
TDE

Team*

Tecloftalam

Tedion

Teflubenzuron

Terbacil

Terbufenpyrad

Terbuthylazine

Terephthaloyl chloride

Terraguard*

Terraneb*

Terrazan*

Terr-0-Gas*

Tersan* 75

Tetrabromo-*o*-cresol

Tetrabromoethane

Tetrachlorobenzene

2,3,7,8-Tetrachlorodibenzo-*p*-dioxin

Tetrachlorodinitroethane

Tetrachlorodiphenylethane

Tetrachloroethane

Tetrachloroethylene

Tetrachloroisophthalonitrile

Tetrachloromethane

Tetrachloronaphthalene

Tetrachloronitroanisole

Tetrachlorophenol

Tetrachloropropyl ether

Tetrachlorothiophene

Tetrachlorovinphos

Tetraconazole

Tetradifon

Tetrafluoroethane

Tetrafluoroethylene (Perfluoroethylene, TFE)

Tetrafluorohydrazine

Tetrafluoromethane

Tetramethylammonium chlorodibromide

Thalonex*

Thiobencarb

2,2'-Thiobis (chlorophenol)

Thiocarbonyl chloride

Thiodan*

Thionate*

Thionyl chloride

Thiophosgene

Thiophosphoryl chloride

TIBA

Titanocene dichloride

Tobosa*

Tok*

Tokuthion*

Tolclofos-methyl

Toxaphene

Toxaphene with methyl parathion

2,4,5-TP

Tralex*

Tralomethrin

Transmix*

Treflan*

Triadimefon

Trialin*

Triazine pesticide (Dyrene*, Anilazine)
Triazoxide
Tribromoacetaldehyde
Tribromoacetic acid
Tribromo-*tert*-butyl alcohol
Tribromoethanol
Tribromomethane
Tribromophenol
Tribromopropane
3,4,5-Tribromosalicylanilide
Tricamba
Trichlorfon (Trichlorphon)
Trichloroacetaldehyde
Trichloroacetic acid
Trichloroacetonitrile
Trichloroacetyl chloride
S-2,3,3-Trichloroallyl-*N,N*-diisopropylthiolcarbamate
Trichloroanisole
Trichlorobenzene(s)
Trichlorobenzoic acid
Trichlorobenzyl chloride
2,3,6-Trichlorobenzyloxypropanol
Trichlorobutene
Trichloroethane (Various forms)
Trichloroethanol
Trichloroethylene (Trichloroethene)
Trichlorofluoromethane (Fluorocarbon-11)
Trichloroisocyanuric acid
Trichloromelamine
Trichloromethane

Trichloromethyl chloroformate

Trichloromethyl ether

Trichloromethylphosphonic acid

Trichloromethylsulfenyl chloride

Trichloronaphthalene (Halowax)

Trichloronate

Trichloronitromethane

Trichloronitrosomethane

Trichlorophenol (Various forms)

2,4,6-Trichlorophenol

Trichlorophenoxyacetic acid

(2,4,5-Trichlorophenoxy)propionic acid (Silvex)

2,4,5-Trichlorophenyl acetate

Trichloropropane

1,2,3-Trichloropropane

Trichlorosilane

Trichloro-*s*-triazinetrione, dry (Trichloroisocyanuric acid)

(*mono*)-(Trichloro)-*tetra*(monopotassium dichloro)*penta-s*-triazinetri-one, dry

Trichlorotrifluoroacetone

1,1,2-Trichloro-1,2,2-trifluoroethane

Triclopyr

Triclopyr, triethylamine salt

Trietazine

Tri-Ethane*

Triflumizole

Trifluoroacetic acid

Trifluoroacetyl chloride

Trifluorobromoethane (Halon 1301)

Trifluorobromomethane

Trifluorochloroethylene

Trifluoroethane

Trifluoromethane

Trifluoromethylaniline (Various forms)

Trifluoromethylbenzene

Trifluoronitrosomethane

Trifluralin

Triforine

Triiodobenzoic acid

Triiodomethane

Trimedlure

Trimethylacetyl chloride

Trimethylchlorosilane

Trimeturon

Trimidal (Triminol*)

Trinatox D*

Tris(2-chloroethyl)amine

Tris(2,3-dibromopropyl)phosphate

Tris(2,3-dichloropropyl)phosphate

Tri-Scept*

Tritac*

Tritox*

d-Tubocurarine chloride

Uniconazole-P (Uniconazole)

Valeryl chloride

Valium*

Vapona* (Vaponite*)

VC

Vegadex*

Vengador*

Verdict*
Victoria Green
Vigil*
Vinclozolin
Vinyl bromide
Vinyl chloride (VC)
Vinyl chloroacetate
Vinyl-2-chloroethyl ether
Vinyl fluoride
Vinylidene chloride
Vinylidene fluoride
Vinyl trichloride
Vinyl trichlorosilane
Voltage
Vorlan*
Vulklor*
Wax, chloronaphthalene
Xylyl bromide
Xylyl chloride
Zeidane
Zeta-Cypermethrin
Zinc trichlorophenate
Zirconocene dichloride
Zobar*

RGN 18 ISOCYANATES, ISOTHIOCYANATES, AND THIOCYANATES (ALL ISOMERS)

Isocyanates, isothiocyanates, and thiocyanates are poisonous, reactive, highly flammable, and tend to polymerize. The compounds can be easily ignited by flames, sparks, or heat. Vapors form explosive mixtures with air and are explosion hazards indoors, outdoors, and in sewers and confined spaces. Some vapors may travel to source of ignition and flash back. These compounds will react with water (water-sensitive under slightly acidic and/or basic conditions) releasing toxic, flammable, and/or corrosive gases. The compounds are extremely reactive with acids and many other chemicals. Containers may explode when heated or if contaminated with water. Inhalation and contact with these chemicals may cause irritation, burns, injury, or death. Trained and experienced individuals should handle these chemicals properly.

Allyl isocyanate

Allyl isothiocyanate

Benzyl isothiocyanate

Benzyl thiocyanate

β-Butoxy-β'-thiocyanodiethyl ether

Butyl isocyanate

3-Chloro-4-methylphenyl isocyanate

Chlorophenyl isocyanate

Cyclohexyl isocyanate

Dianisidine diisocyanate

Dichlorophenyl isocyanate(s)

Diphenylmethane-4,4'-diisocyanate (MDI)

Ethoxycarbonyl isothiocyanate

Ethyl isocyanate

Ethyl mustard oil (Ethylthiocarbimide)

Hexamethylene diisocyanate

Isobornyl thiocyanoacetate

Isobutyl isocyanate

Isocyanate solution, n.o.s.

Isocyanates, n.o.s.

Isocyanatobenzotrifluorides

Isophorone diisocyanate (IPDI)

Isopropyl isocyanate

Lethane*

Methoxymethyl isocyanate

Methylene bis(4-cyclohexylisocyanate)

Methylene diisocyanate

Methyl isocyanate

Methyl isothiocyanate

Mustard Lewisite

Mustard oil

Pharaoh's serpent eggs (Mercuric thiocyanate)

Phenyl isocyanate

Phenyl mustard oil

Polyphenyl polymethylisocyanate

Propyl isocyanate

Toluene diisocyanate (Toluene-2,4-diisocyanate, Toluene-2,6-diiso-
     cyanate)

Trapex*

RGN 19 <u>KETONES</u> (ALL ISOMERS)

Ketones (low molecular weight) tend to be highly flammable and can be ignited by heat, flame, or sparks. Some larger compounds are less volatile and combustible. Vapors may form explosive mixtures indoors, outdoors, and in sewers and enclosed areas. Many vapors travel to the source of ignition and flashback. Containers may explode when heated. Fire may produce irritating, corrosive, and/or toxic gases.

Acetol

Acetone

Acetone oils

Acetonylacetone

Acetophenone

Acetylacetone

Acetyl acetone peroxide

Acetylmethylcarbinol (Acetoin)

ACN (ACNQ)

Allethrin

Allylacetone

Amyl methyl ketone

Anisylacetone

BEK

Benzophenone

Benzoquinone

Bis(dimethylamino)benzophenone

Bromoacetone

Bromobenzoyl acetanilide

Bromomethylethyl ketone

Butanedione

Butanone

Buten-2-one

Butopyronoxyl

Butyl ethyl ketone

Butyl mesityl oxide

Camphor

2-Carbethoxycyclopentanone

Castellan*

Chlordecone

Chloroacetone

Chloroacetophenone

Chlorophacinone

Chrysanthemte compounds (Pyrethrins)

Cinerin I

Cinerin II

Clethodim

Coumafène (Warfarin)

Coumafuryl

Coumatetralyl

Curalin*

Cyclohexanone

Cyclopentanone

Dehydroacetic acid (DHA)

Diacetone alcohol

Diacetone alcohol peroxide(s)

Diacetyl

Diazepam

1,2-Dibromobutan-3-one

Di-*tert*-butylbenzoquinone

Diclone

Dichloroacetone

Dichloro-5,6-dicyanobenzoquinone

Dichloro-*s*-triazine-2,4,6-trione

Diclomezine

Diethyl ketone

Diisobutyl ketone

Diisopropyl ketone

2,6-Dimethyl-4-heptanone

Dimethylketone

Diphacinone (Daphacin)

Diphenadione

Dipropyl ketone

Drazoxolon

2-[1-(Ethoxyimino)butyl]-5-[2-(ethylthio)propyl]-3-hydroxy-2-cyclohexen-1-one

Ethyl amyl ketone

Ethyl butyl ketone

Ethyl methyl ketone

Ethyl propyl ketone

Exxpel 2*

Fluoroacetophenone

Fluquinconazole

Fumarin

Grasp*
Griseofulvin
2-HAP
Heptanone(s)
Hexanone(s)
Hexazinone
Hydroxyacetone
Hydroxyacetophenone
Hydroxypropanone
Isophorone
Isophoronediamine
Isophorone diisocyanate
Isopropylacetone
Kepone®
Ketones, liquid, n.o.s.
Martonite
Mesityl oxide
4-Methoxy-4-methylpentan-2-one
Methyl acetone
Methylamyl ketone
Methylbromoacetone
Methylbutan-2-one
Methyl butyl ketone (various forms)
Methylcyclohexanone (various forms)
Methyl ethyl ketone (MEK)
5- Methyl-3-heptanone
5- Methylhexan-2-one
Methyl isoamyl ketone
Methyl isobutyl ketone
Methyl isobutyl ketone peroxide

Methyl isopropenyl ketone

Methyl isopropyl ketone

Methyl propyl ketone

N-Methyl-2-pyrrolidone

Methyl vinyl ketone

Metribuzin

MIBK

Michler's ketone

Mikado*

Mogeton G*

Monochloroacetone (Chloroacetone)

MPK

Musk ketone (Perfume)

Naramycin

4-(N-Nitrosomethylamino)-1-(3-pyridyl)-1-butanone

NNK

Nonanone

Octanone

Ornalin*

Oxadiazon

2,4-Pentanedione (Pentane-2,4-dione, Pentan-2,4-dione)

Pentanone(s)

Phenacyl bromide

Phenacyl chloride

Phenacyl fluoride

Phenylbutazone

Phenylchloromethyl ketone

Phenyl methyl ketone

Pindone (2-Pivaloyl-1,3-indandione)

Prednisolone

Pynamin*
Pyrazoxyfen
Pyrellin*
Pyrethrins I and II (Pyrethrines)
Pyrethroid pesticides
Pyrethrum
Pyridaben
Pyroquilon
Pyruvic alcohol
Quinoclamine
Quinone
Ronilan*
Rotenone
Saxitoxin
Select*
Sethoxydim
Solvenol*
Sulcotrione
Tetranactin
Thiofanox (Thiofanocarb)
Thujone
Tomarin
Tralkoxydim
Triadimefon
1,3,5-Triglycidyl-s-triazinetrione
Trimethadione
Trimethyl-2-cyclohexene-1-one
Valium*
Valone
Vinclozolin

Vinyl methyl ketone
Vinyl-2-pyrrolidine
Vorlan*
Vulklor*
Warfarin
Xanthone
Zingerone
Zoocoumarin

RGN 20 MERCAPTANS, THIOLS, AND ORGANIC SULFIDES (ALL ISOMERS)

Many of the compounds are stable at room temperature, but tend to decompose at higher temperatures. The compounds tend to be odorous, toxic, and flammable. Many of these compounds tend to be highly flammable and are an explosion risk. Toxic vapors of hydrogen sulfide, organic sulfides, sulfur dioxides, carbon disulfide, and additional sulfur and sulfur-nitrogen compounds can form during a fire or during decomposition of many of these materials. Vapors may travel to the source of ignition and flash back and are an explosion hazard indoors, outdoors, or in sewers and enclosed spaces. Containers may explode when heated. Avoid breathing any gaseous vapors from these chemicals and their decomposition and/or reaction products. Some compounds are lachrymators, while others can be easily absorbed by the skin. Several sulfur-based poisons and war gases (types of mustard gases) can be effectively rendered nonharmful or decontaminated by *strong* bleaching agents or oxidizing agents such as chlorox, chloroamines, etc. Trained and experienced individuals should handle these chemicals properly.

Aimcosystox*

Alanycarb

Aldicarb (Aldicarbe)

Allyl propyl disulfide

Allyl sulfide

Ammonium thioglycolate

Amyl mercaptan

BAL
Beam*
Benzoepin
Benzothiazole
Benzothiazolyl disulfide
Benzyl disulfide
Benzyl mercaptan
Benzyl sulfide
Benzylthiol
Bim*
Bis-(2-chloroethyl)sulfide
Bis-(2-chloroethyl)sulphide
Bis(3-nitrophenyl)disulfide
Bis(tetrachloroethyl)disulfide
Bithionol
Blas-T*
Bromlost
Bromothiophenol
Butanethiol
Butoxycarboxim
Butyl mercaptan
Butyl sulfide
Captab
Captafol*
Carbathiin
Carbosulfan
Carboxin (Carboxine)
Cetyl mercaptan
Chinomethionat
Chlorbenside

3-(*p*-Chlorophenyl)-5-methylrhodanine

Chloropropyl mercaptan

Chlorothiophenol

Clethodim

Croneton*

Cyclohexanethiol

Cyclohexyl mercaptan

Diacetonyl sulfide

Di-*tert*-amyl disulfide

Diamyl sulfide

Dibenzyl disulfide

Dibromodiethyl sulfide

Di-*tert*-butyl disulfide

Di-*tert*-butyl sulfide

Dibutylxanthogen disulfide

Dicarboximides (Sulfenimides)

Dichlorodiethyl sulfide (Mustard gas)

Dictran*

Diethylene disulfide

Diethyl sulfide

2,2'-Dihydroxy-5,5'-difluorodiphenyl sulfide

Dihydroxyethyl sulfide (Thiodiglycol)

Diiododiethyl sulfide

Diisopropyl dixanthogen

Dimension*

Dimepax*

Dimercaptopropanol (BAL)

Dimethametryn

Dimethyl disulfide

Dimethyl sulfide

3,5-Dimethyltetrahydro-1,3,5(2H)thiadiazine-2-thione

Dimyristyl sulfide

Dipicryl sulfide

Dipyridylethyl sulfide

Disulfiram (TTD, TETD)

1,4-Dithane*

Dithianone

Dithiobis(benzothiazole) (MBTS)

Dithiobiuret

Dithiocarbamates, pesticide, n.o.s.

Dithiocarbamic acid

Dithiodimorpholine

Dithiopyr

Divinyl sulfide

Dixanthogen

Dodecyl mercaptan (DDM)

Endosulfan

Ethanedithiol

Ethanethiol

Ethiofencarb

2-[1-(Ethoxyimino)butyl]-5-[2-(ethylthio)propyl]-3-hydroxy-2-cyclohexen-1-one

Ethyl mercaptan

Ethyl sulfide (Diethyl sulfide)

Ethylthiodemeton (M-74, Disulfoton)

Ethyl thioethanol

Ethyl tuex*

Etridiazole

Evisect*

Fenamiphos

Fentichlor (Novex)
Fluorbenside
Fopel
Hexyl mercaptan
Isoamyl mercaptan
Isobutyl mercaptan
Isooctyl thioglycolate
Isopropyl mercaptan
Isoprothiolane
Kayaphos*
Lannate*
Lauryl mercaptan
Luxafan*
MBTS
MCPA-thioethyl
Mercaptans and Organic Sulfides, n.o.s.
Mercaptoacetic acid
Mercaptobenzothiazole (MBT)
Mercaptodimethur
Mercaptoethanol
Mercaptopropionic acid
Mesurol*
Methanethiol
Methiocarb
Methomyl
2-Methyl-2-butanethiol
2-Methyl-2-heptanethiol
Methyl mercaptan
2-Methyl-2-propanethiol
6-Methyl-2,3-quinoxalinedithiol cyclic carbonate

Methyl sulfide

(Methylthio)aniline

4-(Methylthio)-3,5-dimethylphenyl-*n*-methylcarbamate

Morestan*

MSR2*

Mustard gas

Naphtyl mercaptan

Nemacur*

tert-Octyl mercaptan

Onic*

Organic Sulfide, n.o.s.

Oxamyl

Oxythioquinox

Pentanethiol

Perchloromethyl mercaptan

Phenothiol

Phenothiazine

Phenyl mercaptan

Phorate

Phosfolan (Phospholan)

Polysulfide polymer

Prometryn

Propafos

Propanethiol(s)

Propaphos

2-Propene-1-thiol

Propyl mercaptan

Pyridaben

Pyridate

Rhodanine

Sanacarb*

Select*

Spike*

Sulfur mustard

Sunter*

Synthetic rubber decomposition products (Containing sulfur)

Tebuthiuron (Tebusan*)

Temik*

Terbufos

Terbutryn (Ternit*)

Tersan* 75

TETD

Tetrabutylthiuram disulfide

Tetradecyl thiol

Tetraethylthiuram disulfide (TTD, TETD)

Tetrahydrothiophene

Tetraisopropylthiuram disulfide

Tetrapropylthiuram disulfide

Tetrone*

Thimet*

Thioacetamide

Thioacetic acid

Thiobenzoic acid

Thiodicarb

4,4'-Thiobis(6-*tert*-butyl-*m*-cresol)

2,2'-Thiobis(chlorophenol)

Thiocarbonyl chloride

Thiocarboxime

Thioclam hydrogen oxalate (Thioclam)

Thiodan*

Thiodicarb

Thiodiglycol

Thiodiphenol (TDP)

Thiodipropionic acid

β,β-Thiodipropionitrile

Thiofanox (Thiofanocarb)

Thioglycol

Thioglycolic acid (Mercaptoacetic acid)

Thiolactic acid

Thiols, aliphatic

Thiometon

Thionate*

Thiophene

Thiophenol

Thiophosgene

Thioquinox

Thiosalicylic acid

Thiosemicarbazide

Thiotepa or tris(1-Aziridinyl)phosphine sulfide

Thioxylenol

Thiram (Thirame, Thiuram, TMTD)

Thiramad*

Thiuram

Timet

TMTD

Toluenethiol

Trichloromethylsulfenyl chloride

Tricyclazole

Triethylenethiophosphoramide

TTD

Vydate G*

Zinc-2-mercaptobenzothiazole

RGN 21 METALS, NONMETALS, AMALGAMS, AND SPECIFIC ALLOYS

Metals, amalgams, and specific alloys react with water (moisture), and/or air. Some metals ignite spontaneously while others react violently (explosively) in air, oxygen, and/or water. Some fine powders and dusts form flammable and explosive mixtures in air. A number of metals can be classified as *pyrophoric* at room temperature. In general, these metals are a dangerous fire and explosion risk in the presence of oxidizing materials and under a variety of conditions. Many metals evolve hydrogen gas when in contact with moisture or water. Lithium reacts explosively with nitrogen in moist air. After reaction with water or air, the toxicity level of several metals tends to increase due to increases in metal solubility and changes in the oxidation state. These chemicals react with moist tissue and/or skin. Experienced individuals should handle these chemicals properly.

Alkali metal(s), n.o.s.

Alkali metal amalgams, n.o.s. (Liquid or Solid)

Alkaline-earth metal alcoholates, n.o.s.

Alkaline-earth metal alloys (Amalgams)

Aluminum powder (Pyrophoric; Fine powder)

Aluminum silicon powder (Uncoated)

Barium (Pyrophoric; powder)

Barium alloys (Pyrophoric)

Boron (Dust)

Calcium (Metal and Alloys; Pyrophoric)

Cerium (Turnings or gritty powder)

Cerium-iron alloys (Potential pyrophoric)

Cesium (Caesium)

Europium (Powder, Highly reactive)

Ferrocerium (Alloy of iron and misch metal)

Hafnium powder (Dry or wet)

Lanthanum (Powder)

Lithium

Magnesium (Powder, Granules, Turnings, Ribbons, Pellets, Flakes, etc.)

Magnesium alloys (With more than 50% Mg in pellets, turnings or ribbon)

Magnesium alloy powder

Metal powder (Self-heating; n.o.s.)

Metals, n.o.s. (Water reactive and/or Air sensitive)

Metal catalyst, dry (Water reactive and/or Air sensitive)

Potassium (Metal, Alloys, Liquid; Ignites readily in air)

Potassium-40, 42 (Radioactive metal)

Potassium-sodium alloy (Ignites in humid air)

Phosphorus (Amorphous red; Oxygen and water-vapor sensitive)

Phosphorus (White; Ignites spontaneously in air)

Radium-226 (Air sensitive and water reactive)

Raney nickel (Air sensitive; Store under water or alcohol)

Rubidium

Sodium

Sodium amalgam (Sodium-mercury amalgam; Water sensitive)

Sodium-lead alloy (Moisture sensitive)

Sodium-potassium alloy (NaK; Ignites in humid air)

Strontium (Powder)

Thallium

Thorium (Pyrophoric metal powder)

Thorium alloys (Some ignite spontaneously)

Titanium (Powder, Dry; May ignite; Fire and explosion risk)

Uranium (Powder; Pyrophoric metal; Radioactive risk)

Zinc (Dust, Powder, Dry or Damp; Air sensitive)

Zinc foil (Ignites in moisture)

Zinc dross (Water-reactive)

RGN 22 METALS AND ALLOYS—DUSTS, POWDERS, FUME, AND VAPORS

In acid, some metals are flammable while a few metals form hydrogen gas with the potential for explosion. Additional metal dusts and/or powders are flammable. Some metals are very toxic while others display limited toxicity. Specific metals, such as mercury and uranium, are highly toxic by skin absorption and inhalation of fumes and vapor. A few metal dusts are poisons (toxic), animal carcinogens, and/or human carcinogens. Avoid inhalation of these metals and alloys as dusts and fine powders.

Aluminum silicon powder (Flammable)

Antimony fume, fine powder

Arsenic (Carcinogen and Mutagen)

Arsenical dust (Carcinogen)

Babbitt metal dust (Toxic by inhalation)

Bagasse dust (Flammable)

Beryllium dust or fine powder (Toxic; Carcinogen)

Beryllium-copper alloy dust (Potential carcinogen)

Bismuth powder (Flammable)

Brass as fine powder (Flammable)

Bronze powder (Flammable)

Cadmium dust, fume, powder (Carcinogen; Poison; Toxic; Flammable)

Carbon black or carbon dust (Toxic; Flammable)

Cobalt dust (Animal carcinogen; Flammable)

Copper in finely divided form (Flammable)

Copper dusts, mists, fume (Toxic)

Erbium in finely divided form (Flammable)

Ferrosilicon (Flammable with 30–90% silicon and evolves gases when damp)

Ferrovanadium dust (Moderate fire risk; Toxic)

Graphite dust or black lead (Fire risk, Respirable dust)

Indium (Toxic)

Iron dust and fine particles (Flammable and explosion risk)

Lead fumes, dust (Toxic; Cumulative poison)

Manganese (Dust, Powder; Toxic; Flammable)

Mercury (Fume or vapor, Absorbed; Toxic)

Metals, n.o.s.

Molybdenum (Dust , Powder; Toxic; Flammable)

Nickel (Dust, Fume; Toxic; Flammable; Carcinogen)

Nickel (Catalyst, dry)

Osmium (Toxic)

Platinum (Powder; Flammable)

Uranium (Powder; Flammable; Toxic; Radioactive)

Quicksilver or Mercury (Fume or vapor, Absorbed; Toxic)

Rhenium (Powder; Flammable)

Rhodium (Powder; Flammable)

Silicon (Powder; Flammable)

Silver (Toxic)

Sulfur (Finely divided form; Fire and explosion risk)

Tantalum (Dust or Powder; Flammable; Toxic)

Thulium (Dust; Fire risk)

Titanium (Wetted powders; Dangerous fire and explosion risk; Toxic)

Titanium sponge (Granules, Powders; Dangerous fire and explosion risk; Toxic)

Tungsten (Finely divided form; Fire risk)

Welding fumes (Toxic metals)

Yttrium (Finely divided form, Flammable; Toxic)

Zirconium (Dust, Powder, Borings, Shavings, Liquid suspension, Scrap, etc.; Wet or Dry; Fire and explosion risk; Suspected carcinogen)

RGN 23 RADIOACTIVE POISONS, ISOTOPES, METALS, IONS, ELEMENTS, AND COMPOUNDS

Some nonmetals, metals, and compounds are radioactive. Some radioactive materials are commonly identified as radioactive poisons or toxic materials. These radioactive substances have a variety of applications as elements (metals and nonmetals), compounds, and ions. Depending on the application, many of the species can exist as gases, solids, or in aqueous and/or nonaqueous solution. Some radioactive isotopes, such as lead, are stable and are the products of disintegration of natural radioactive elements. Other isotopes exist, but have limited applications.

Actinium (Radioactive poison)

Americium (Radioactive poison)

Antimony-124, 125 (Radioactive poison)

Barium-137

Boron-10 (Nonradioactive isotope)

Calcium-45

Carbon-11, 13, 14

Carnotite (Radioactive poison, Uranium vanadate-type complex)

Cerium-141 (Radioactive poison)

Cesium-137 (Radioactive poison)

Chlorine-35, 36, 37 (Radioactive gas and compounds, Poisonous gas)

Chromium-51 (Radioactive poison)

Cobalt-57, 58, 60 (Radioactive poison)

Curium (Synthetic radioactive element)

Deuterium (Heavy hydrogen, Gas; D-2)

Europium (Radioactive isotope)

Gold-198

Heavy water (Deuterium oxide and tritium oxide)

Helium-3

Iodine-131

Iridium-192

Iron-55, 59 (Toxic material)

Krypton-86 (Gas)

Lead (Four stable isotopes)

Lutetium-176 (Radioactive isotope)

Mercury (Radioactive isotopes)

Neptunium (Radioactive poison)

Neptunium dioxide (Radioactive poison)

Nitrogen-15 (Radioactive isotope)

Oxygen-17, 18 (Radioactive element, Heavy oxygen, Gas)

Phosphorus-32 (Radioactive compounds and/or ions)

Pitchblende (Radioactive material)

Platinum (Radioactive isotopes)

Plutonium (Extremely Toxic, Radiotoxic element, Powerful carcinogen)

Polonium (Radioactive poison, Radiotoxic; Cigarette smoke)

Potassium-40, 42 (Metal and/or Compounds; Radioactive isotopes)

Promethium-147 (Radioactive poison)

Radium-226 (Radioactive toxic element, Destructive)

Radium-226 (Radioactive compounds)

Radon-222 (Radioactive toxic gas; Human carcinogen)

Scandium (Artificial radioactive isotope)

Sodium diuranate (Radioactive compound)

Sodium phosphate (P-32, Radioactive form)

Strontium-89, 90 (Radioactive poison)

Sulfur-35 (Radioactive poison)

Technetium-99 (Radioactive)

Thorium-232 (Thorium-232 nitrate; Radioactive compound and/or element)

Thulium-170 (Radioactive source)

Tritium-3 (Radiotoxic gas)

Triuranium octoxide (Radioactive poison)

Uranium-233,234,235,238 (Isotopes, Metal, Alloys, Powder;)

Uranium-233,234,235,238 compounds, n.o.s. (Radioactive risk or Radioactive poison)

Uranium dicarbide (Radioactive risk)

Uranium dioxide (Radioactive risk)

Uranium hexafluoride (Radioactive risk)

Uranium hydride (Radioactive risk)

Uranium nitrate (Uranyl nitrate; Radioactive risk)

Uranium sulfide (Radioactive risk)

Uranium monocarbide (Radioactive poison)

Uranium tetrafluoride (Green salt; Radioactive poison)

Uranium yellow (Radioactive compound)

Uranyl acetate (Radioactive risk)

Uranyl nitrate hexahydrate (Radioactive risk)

Yellow salt (Radioactive risk)

Zinc-65 (Radioactive poison)

Zirconium-95 (Radioactive poison: Compounds usually)

RGN 24 INORGANIC COMPOUNDS, METALS, ORGANOMETALLIC SUBSTANCES, AND SPECIAL COMPOUNDS

Many of these compounds are noncombustible. However, most compounds are toxic and solubilize in water relatively easily. Some compounds are corrosive to skin and tissue and are highly toxic by skin absorption, inhalation, and ingestion. A few compounds are toxic by inhalation of dust particles or mists. A number of compounds are known poisons and/or carcinogens. Additional dusts and powders are nuisance particles. Some of the compounds are oxidizing agents. Avoid breathing vapors from these chemicals or the vapors from decomposition products of the chemicals.

Acetomeroctol

Adamsite

Aluminum arsenide

Aluminum distearate

Aluminum fluosilicate

Aluminum monostearate

Aluminum oxide, dust

Aluminum phosphate

Aluminum resinate

Aluminum stearate

AMA*

Amine methanearsonate (AMA)

Ammonium arsenate

Ammonium chromate

Ammonium dichromate

Ammonium hexanitrocobaltate

Ammonium hexachloroplatinate

Ammonium methanearsonate

Ammonium metavanadate

Ammonium molybdate

Ammonium permanganate

Ammonium polyvanadate

Ammonium tetrachromate

Ammonium tetraperoxychromate

Ammonium trichromate

Antimony chloride

Antimony compound, n.o.s.

Antimony fluoride

Antimony hydride (Stibine)

Antimony lactate

Antimony nitride

Antimony oxide

Antimony oxychloride

Antimony pentachloride

Antimony pentafluoride

Antimony pentasulfide

Antimony pentoxide

Antimony perchlorate

Antimony potassium tartrate

Antimony salt (DeHaens salt)

Antimony sulfate

Antimony sulfide (Antimony sulphide)

Antimony tribromide

Antimony trichloride

Antimony trifluoride

Antimony triiodide

Antimony trioxide

Antimony trisulfate

Antimony trisulfide (Antimony trisulphide)

Antimony trivinyl

Antimony yellow (Naples yellow)

Arsanilic acid

Arsenic acid

Arsenical pesticide

Arsenic bromide

Arsenic chloride

Arsenic compound, n.o.s.

Arsenic disulfide

Arsenic fluoride

Arsenic iodide

Arsenic oxide

Arsenic pentafluoride

Arsenic pentaselenide

Arsenic pentasulfide

Arsenic pentoxide

Arsenic sulfide

Arsenic tribromide

Arsenic trichloride

Arsenic trifluoride

Arsenic triiodide

Arsenic trioxide
Arsenic trisulfide
Arsenious oxide
Arsine
Barium acetate
Barium azide
Barium borotungstate
Barium bromate
Barium bromide
Barium carbide
Barium carbonate
Barium chlorate
Barium chloride
Barium chromate
Barium citrate
Bariun cyanide
Barium cyanoplatinite (Barium platinum cyanide)
Barium dichromate
Barium dithionate (Barium hyposulfate)
Barium ethylsulfate
Barium fluoride
Barium fluosilicate
Barium hydride
Barium hydroxide
Barium hypochlorite
Barium hypophosphite
Barium iodate
Barium iodide
Barium manganate
Barium metaphosphate

Barium molybdate
Barium nitrate
Barium nitrite
Barium oxalate
Barium oxide
Barium perchlorate
Barium permanganate
Barium peroxide
Barium phosphate
Barium potassium chromate
Barium selenate
Barium selenide
Barium silicate
Barium silicide
Barium silicofluoride
Barium stannate
Barium stearate
Barium sulfate
Barium sulfide
Barium sulfite
Barium tartrate
Barium thiocyanate
Barium thiosulfate
BCWL
Benzoylferrocene
Beryl
Beryllium acetate
Beryllium chloride
Beryllium compound, n.o.s.
Beryllium fluoride

Beryllium hydride

Beryllium hydroxide

Beryllium metaphosphate

Beryllium nitrate

Beryllium oxide

Beryllium phosphate

Beryllium potassium sulfate

Beryllium sodium fluoride

Beryllium sulfate

Beryllium sulfate tetrahydrate

Beryllium sulfide

Beryllium tetrahydroborate

Beryllium zinc silicate

Beryl ore

Bismuth chromate

Bismuth ethyl chloride

Bismuthic acid

Bismuth iodide

Bismuth pentafluoride

Bismuth pentoxide

Bismuth stannate

Bismuth sulfide

Bismuth telluride

Bismuth tribromide

Bismuth trichloride

Bismuth triiodide

Bismuth trisulfide

Bis(tri-*n*-butyltin) oxide

Bisulfates (Bisulphates), inorganic, n.o.s.

Bisulfites (Bisulphites), inorganic, n.o.s.

Borane

Borax (Sodium borate, Tincal)

Borax pentahydrate

Bordeaux arsenites

Boric oxide (Boron oxide)

Boron arsenotribromide

Boron bromodiiodide

Boron dibromoiodide

Boron nitride

Boron oxide

Boron phosphide

Boron triazide

Boron tribromide

Boron trichloride

Boron trifluoride

Boron trifluoride, dihydrate

Boron trifluoride acetic acid complex

Boron trifluoride monoethylamine

Boron trifluoride propionic acid complex

Boron triiodide

Boron trisulfide

Butyldichloroarsine

Butylstannic acid

Butyltin chloride

Cacodylic acid

Cadminate*

Cadmium acetate

Cadmium acetylide

Cadmium amide

Cadmium azide

Cadmium borotungstate
Cadmium bromate
Cadmium bromide
Cadmium carbonate
Cadmium chlorate
Cadmium chloride
Cadmium compound, n.o.s.
Cadmium cyanide
Cadmium diethyldithiocarbamate
Cadmium fluoride
Cadmium hydroxide
Cadmium iodate
Cadmium iodide
Cadmium molybdate
Cadmium nitrate
Cadmium nitride
Cadmium oxide
Cadmium phosphate
Cadmium potassium iodide
Cadmium ricinoleate
Cadmium sebacate
Cadmium stearate
Cadmium succinate
Cadmium sulfate
Cadmium sulfide
Cadmium telluride
Cadmium tungstate
Calcium arsenate
Calcium arsenite
Calcium fluosilicate

Calcium hydrogen sulfite

Calcium propanearsonate

Calcium resinate

Calcium selenate

Calcium silicofluoride

Calcium stannate

Calcium stearate dust

Calcium sulfate dust

Calo-Clor*

Calocure*

Calogran*

Calomel (Mercurous chloride)

CCA*

Ceramix*

Ceresan*

Cerous fluoride

Cerous oxalate

Cesium antimonide

Cesium arsenide

Chlorochromic anhydride

Chloromercuriferrocene

Chloroplatinic acid

β-Chlorovinyldichloroarsine (Lewisite)

Chlorovinylmethylchloroarsine

Chromated copper arsenate (CCA)

Chrome alum

Chrome pigment

Chromic acetate

Chromic acid

Chromic anhydride

Chromic chloride

Chromic fluoride

Chromic nitrate

Chromic oxide

Chromic sulfate (Chromic sulphate)

Chromic sulfide

Chromite

Chromium compounds (Hexavalent chromium)

Chromium hexacarbonyl

Chromium naphthenate

Chromium nitrate

Chromium oxychloride

Chromium oxyfluoride

Chromium potassium sulfate (Chrome alum)

Chromium sulfate (Chromium sulphate)

Chromium sulfide (Chromic sulfide)

Chromium trichloride

Chromium trifluoride

Chromium trioxide (Chromic acid)

Chromosulfuric acid (Chromosulphuric acid)

Chromous bromide

Chromous carbonate

Chromous chloride

Chromous fluoride

Chromous oxalate

Chromyl chloride

Chromyl fluoride

Cinnabar

Cobalt acetate (Cobaltous acetate)

Cobalt arsenate (Cobaltous arsenate)

Cobalt bromide (Cobaltous bromide)

Cobalt carbonyl

Cobalt chloride (Cobaltous chloride)

Cobalt chromate (Cobaltous chromate)

Cobalt compounds, inorganic

Cobaltite dust

Cobalt naphthenate

Cobalt nitrate (Cobaltous nitrate)

Cobalt resinate (Cobaltous resinate)

Cobaltocene

Cobaltous chloride

Cobaltous chromate

Cobaltous cyanide

Cobaltous fluoride

Cobaltous formate

Cobaltous nitrate

Cobaltous perchlorate

Cobaltous resinate (Cobalt resinate)

Cobaltous silicofluoride

Cobaltous sulfamate (Cobalt sulfamate)

Cobaltous sulfate (Cobalt sulfate, Cobaltous sulphate)

Cobalt potassium cyanide

Cobalt tetracarbonyl

Cobalt trifluoride (Cobaltic fluoride)

Copper acetoarsenite (Paris green)

Copper acetylide (Cuprous acetylide)

Copper arsenate (Cupric arsenate)

Copper arsenite (Cupric arsenite)

Copper carbonate

Copper chlorate (Cupric chlorate)

Copper chloride (Cupric chloride)
Copper chlorotetrazole
Copper chromate
Copper cyanide (Cupric cyanide)
Copper dihydrazinium sulfate
Copper fluoride (Cupric fluoride)
Copper fluosilicate
Copper hydroxide
Copper metaborate (Cupric borate)
Copper methane arsonate
Copper naphthenate
Copper nitrate (Cupric nitrate)
Copper oxalate
Copper oxide (Red or Black)
Copper oxychloride
Copper oxychloride sulfate
Copper phthalate
Copper 8-quinolate
Copper resinate
Copper ricinoleate
Copper sulfate (Cupric sulfate)
Copper sulfate, ammoniated (Cupric ammonia sulfate)
Copper sulfate, tribasic
Copper zinc chromate
Corrosive sublimate (Mercuric chloride)
Corundum, dust (Aluminum oxide)
Cotton, dust or linters
Cunilate
Cupric acetate
Cupric sulfate (Cupric sulphate)

Cupric sulfate, ammoniated
Cupric tartrate
Cuprobam (Cuprobame)
Cuprous acetylide
Cuprous cyanide
Cuprous iodide
Cuprous mercuric iodide
Cuprous potassium cyanide
Cyano(methylmercury)guanidine
Cyclohexenyltrichlorosilane
Cyhexatin
DAS
DIBAC
DIBAL-H
Dibutyldiphenyl tin
Dibutyltinbis(lauryl) mercaptide
Dibutyltin diacetate
Dibutyltin dichloride
Dibutyltin di-2-ethylhexoate
Dibutyltin dilaurate
Dibutyltin maleate
Dibutyltin oxide
Dibutyltin sulfide
Dichloro-(2-chlorovinyl)arsine
Dichloroethylarsine
Dichloromethylsilane
Dichlorophenarsine hydrochloride
Dichlorovinylchloroarsine
Dichlorovinylmethylarsine
Dicopper(I) acetylide

Dicyclopentadienyl iron
Diethylcadmium
Diethylgermanium dichloride
Diethyl sulfate
Diethylzinc
Dihydroxydiaminomercurobenzene
Diisobutylaluminum chloride
Diisobutylaluminum hydride
Diisopropyl beryllium
Dimethylarsinic acid
Dimethylcadmium
Dimethyldiethoxysilane
Dimethyltin dichloride
Dimethyltin oxide
Diphenylamine chloroarsine (DM)
Diphenylbromoarsine
Diphenylchloroarsine
Diphenylcyanoarsine
Diphenyldichlorosilane
Disodium acetarsenate (Aricyl)
Disodium methylarsonate (DMA)
Disodium trioxosilicate
Disodium trioxosilicate pentahydrate
DM
DSMA
Endothall, sodium salt
Erbium oxalate
Erythrite
Ethoxydimethylsilane
Ethylarsenious oxide

Ethyldichloroarsine
Ethyldichlorosilane
Ethylene chromic oxide
Ethylmercuric acetate
Ethylmercury bromide
Ethylmercury chloride
Ethylmercury 2,3-dihydroxypropyl mercaptide
Ethylmercury iodide
Ethylmercury phosphate
Ethylmercury p-toluenesulfonanilide (EMTS, Ceresan*M)
Ethylmercury sulfate
Ethyl silicate
Ethylzinc
Fentin acetate
Fentin chloride
Fentin hydroxide
Ferric arsenate
Ferric arsenite
Ferric chloride, anhydrous
Ferric chromate
Ferric dichromate
Ferric oxalate
Ferrocene
Ferrous arsenate
Gallium arsenide
Germanium dichloride
Germanium tetrachloride
Greensalt
Hafnium compounds, n.o.s.
Hexaflurate (Potassium hexafluoroarsenate)

Hematite
Hexamethyldisilazane (HMDS)
Hydrargaphen
Hydrides, metal, n.o.s.
Hydrogen selenide gas
Hydrogen telluride gas
Hydroxymercurichlorophenol
Hydroxymercuricresol
Hydroxymercurinitrophenol
4-Hydroxy-3-nitrobenzenearsonic acid
Hydroxyphenylmercuric chloride
Indium antimonide
Indium arsenide
Indium chloride (Indium trichloride)
Indium, compound, n.o.s.
Indium oxide
Indium phosphide
Indium sulfate
Indium telluride
Indium trichloride
Iodine
Iodized oil
Iron arsenate
Iron pentacarbonyl
Isoamyldichloroarsine
L (Lewisite)
Lanthanum arsenide
Lead acetate
Lead alkyl, mixed (In aviation fuel)
Lead antimonate (Naples yellow, Antimony yellow)

Lead arsenate

Lead arsenite

Lead azide

Lead borate

Lead bromate

Lead bromide

Lead carbonate

Lead chloride

Lead chlorite

Lead chromate

Lead compound, soluble, n.o.s.

Lead cyanide

Lead dinitroresorcinate

Lead dioxide

Lead fluoborate

Lead fluoride

Lead fluosilicate

Lead formate

Lead hydroxide

Lead iodide

Lead linoleate

Lead maleate

Lead molybdate

Lead mononitroresorcinate

Lead monoxide (Litharge, Massicot)

Lead naphthalenesulfonate

Lead naphthenate

Lead nitrate

Lead nitride

Lead nitrite

Lead oleate

Lead orthoarsenate

Lead oxide (Yellow, red, brown, black, or hydrated)

Lead perchlorate

Lead peroxide

Lead phosphate (Normal or Dibasic)

Lead phosphite, dibasic

Lead phthalate

Lead resinate

Lead salicylate

Lead selenide, dust

Lead sesquioxide

Lead silicate

Lead silicochromate

Lead sodium thiosulfate

Lead stannate

Lead stearate

Lead styphnate

Lead subacetate

Lead suboxide

Lead sulfate (Lead sulphate)

Lead sulfide

Lead sulfite

Lead tallate

Lead telluride

Lead tetraacetate

Lead thiocyanate

Lead thiosulfate

Lead titanate

Lead trinitroresorcinate

Lead tungstate
Lead vanadate
Lewisite
Litharge
Lithium arsenate
London purple
MAFA (MAF)
Magnesite
Magnesium arsenate
Magnesium arsenite
Magnesium bisulfite solution
Magnesium boron fluoride
Magnesium chloride
Magnesium chromate
Magnesium oxide, fume
Manganese compounds (Inorganic)
Manganese acetate
Manganese arsenate (Manganous arsenate)
Manganese bromide (Manganous bromide)
Manganese carbonyl
Manganese chloride (Manganous chloride)
Manganese cyclopentadienyl tricarbonyl
Manganese methylcyclopentadienyltricarbonyl
Manganese naphthenate
Manganese octoate
Manganese oleate
Manganese resinate
Manganese sulfide
Manganic oxide, dust
Manganous chromate

Mayer's reagent
MEMA
MEMC
Merbromin
Mercuric acetate (Mercury acetate)
Mercuric ammonium chloride
Mercuric arsenate (Mercury arsenate)
Mercuric benzoate (Mercury benzoate)
Mercuric bromide (Mercury bromide)
Mercuric chloride (Mercury chloride)
Mercuric cuprous iodide
Mercuric cyanate
Mercuric cyanide (Mercury cyanide)
Mercuric dimethyldithiocarbamate
Mercuric dioxysulfate
Mercuric fluoride
Mercuric iodide (Mercury iodide)
Mercuric lactate
Mercuric naphthenate (Mercury naphthenate)
Mercuric nitrate (Mercury nitrate)
Mercuric oleate (Mercury oleate)
Mercuric oxide (Mercury oxide)
Mercuric oxycyanide
Mercuric oxycyanide, desensitized
Mercuric phosphate
Mercuric potassium cyanide
Mercuric potassium iodide
Mercuric salicylate
Mercuric silver iodide
Mercuric sodium phenolsulfonate

Mercuric stearate
Mercuric subsulfate
Mercuric sulfate (Mercuric sulphate)
Mercuric sulfide (Red, Black)
Mercuric thiocyanate
Mercurous acetate
Mercurous acetylide
Mercurous bromide
Mercurous chlorate (Mercury chlorate)
Mercurous chloride (Calomel)
Mercurous chromate (Mercury chromate)
Mercurous gluconate
Mercurous iodide
Mercurous nitrate
Mercurous oxide
Mercurous sulfate (Mercurous sulphate)
Mercurous sulfide
Mercury
Mercury (metallic)
Mercury (vapor)
Mercury acetate
Mercury acetylacetone
Mercury, alkyl compounds
Mercury, ammoniated
Mercury ammonium chloride
Mercury, aryl compounds
Mercury based pesticides
Mercury benzoate
Mercury bisulfate (Mercury bisulphate)
Mercury bromide

Mercury compound, n.o.s.

Mercury chlorate

Mercury chromate

Mercury cyanide

Mercury fulminate (Mercuric cyanate)

Mercury gluconate

Mercury iodide

Mercury metal

Mercury nitride (Trimercury dinitride)

Mercury nucleate

Mercury oleate

Mercury oxide

Mercury oxycyanide, desensitized

Mercury potassium iodide

Mercury salicylate

Mercury selenide

Mercury sulfate (Mercury sulphate)

Mercury thiocyanate

Mercusol

Metal alkyl solution, n.o.s.

Metal alkyl halides, n.o.s.

Metal alkyls, n.o.s.

Metal aryl halides, n.o.s.

Metal aryl hydrides, n.o.s.

Metal aryls, n.o.s.

Metal carbonyls, n.o.s.

Metals, n.o.s. (Toxic)

Metal Compounds, n.o.s. (Toxic)

Methoxyethylmercuric acetate

Methoxyethylmercuric chloride

Methylcyclopentadienyl manganese tricarbonyl
Methyl dichloroarsine
Methylmercury acetate
Methylmercury bromide
Methylmercury chloride
Methylmercury cyanide
Methylmercury-2,3-dihydroxypropylmercaptide
Methylmercury hydroxide
Methylmercury iodide
Methylmercury pentachlorophenate
Methylmercury propionate
Methylmercury quinolinolate
Methyl silicate
Misch metal
MLA
Molybdate orange
Molybdenum anhydride
Molybdenum compounds, insoluble and soluble
Molybdenum dioxide
Molybdenum disilicide
Molybdenum hexacarbonyl
Molybdenum naphthalene
Molybdenum pentachloride
Molybdenum sesquioxide
Molybdenum sulfide (Molybdenum disulfide)
Molybdenum trioxide
Molybdic acid
Monosodium acid methanearsonate
Monosodium methanearsonate
Morosodren*

MSMA

Napalm

Neptunium dioxide

Nessler's reagent

New Mel*

Niccolite

Nickel acetate

Nickel ammonium chloride

Nickel ammonium sulfate (Nickel ammonium sulphate)

Nickel antimonide

Nickel arsenate (Nickelous arsenate)

Nickel arsenite (Nickelous arsenite)

Nickel bromide

Nickel carbonate

Nickel carbonyl (Nickel tetracarbonyl; Human carcinogen)

Nickel chloride (Nickelous chloride)

Nickel compounds, insoluble (Lung cancer)

Nickel cyanide

Nickel formate

Nickel hydroxide (Nickelous hydroxide)

Nickel iodide

Nickel nitrate (Nickelous nitrate)

Nickel nitrite

Nickelocene (Human carcinogen)

Nickel oxide

Nickel selenide

Nickel stannate

Nickel subsulfide (Lung cancer)

Nickel sulfate (Confirmed human carcinogen)

Nickel tetracarbonyl (Nickel carbonyl; Human carcinogen)

Niobium potassium oxyfluoride

Nitromersol

Organoarsenic compounds, n.o.s.

Organometallic compounds, n.o.s.

Organotin compound, n.o.s.

Organotin pesticide, n.o.s.

Osmium amine nitrate

Osmium amine perchlorate

Osmium tetroxide (Osmic acid)

Oxalates, metal, water soluble

Palladium nitrate

Paris green

Perlite

Pharaoh's serpent eggs (Mercuric thiocyanate)

Phenarsazine chloride

Phenmad*

Phenylarsonic acid

Phenyldichloroarsine

Phenyllithium

Phenylmercuric acetate

Phenylmercuric benzoate

Phenylmercuric borate

Phenylmercuric chloride

Phenylmercuric compound, n.o.s

Phenylmercuric hydroxide

Phenylmercuric lactate

Phenylmercuric naphthenate

Phenylmercuric nitrate

Phenylmercuric propionate

Phenylmercuriethanolammonium acetate

Phenylmercuritriethanolammonium lactate
Phenylmercury formamide
Phenylmercury salicylate
Phenylmercury urea
Platinium compounds, soluble
Platinium chloride (Platinic chloride)
Platinic sal ammoniac
Plictran* (Cyhexatin)
PMA
Potassium arsenate
Potassium arsenite
Potassium binoxalate (Sorrel salt)
Potassium bisulfite solution
Potassium bromide
Potassium chloroplatinate
Potassium chromate
Potassium dichromate (Potassium bichromate)
Potassium hexafluorophosphate
Potassium manganate
Potassium metavanadate
Potassium monoxide
Potassium osmate
Potassium oxalate
Potassium permanganate
Potassium selenate
Potassium selenite
Potassium stannate
Potassium sulfocarbonate
Potassium thiocyanate
Potassium undecylenate (At high concentrations)

Potassium xanthate

Propineb

Quicksilver

Quinsol (Quinosol)

Red arsenic

Red lead

Rhodium chloride

Rhodium compound(s), n.o.s.

Rohrbach solution

Salicylated mercury

Scheele's green

Selenate (s)

Selenious acid

Selenites

Selenium compound, n.o.s.

Selenium diethyldithiocarbamate

Selenium dioxide

Selenium fluoride

Selenium hexafluoride

Selenium oxide

Selenium oxychloride

Selenous acid (Selenious acid)

Semesan*

Semesan Bel*

Silica (Dust, Fume or Fused)

Silicate (Dust)

Silicon carbide

Silver acetate

Silver acetylide

Silver arsenite

Silver azide
Silver chlorate
Silver chloride
Silver chromate
Silver compound, n.o.s.
Silver cyanide
Silver dichromate
Silver fluoride
Silver methylarsonate
Silver nitrate
Silver nitride
Silver nitrite
Silver oxide
Silver perchlorate
Silver permanganate
Silver peroxide
Silver styphnate
Silver sulfate
Silver sulfide
Silver tetrazene
Silver trinitroresorcinate
Sisal dust
Sneezing gas
Sodium ammonium vanadate
Sodium antimonate
Sodium arsanilate
Sodium arsenate
Sodium arsenite
Sodium bichromate
Sodium bisulfate solution

Sodium bisulfite
Sodium borate
Sodium bromide
Sodium cacodylate
Sodium chromate
Sodium dichromate
Sodium dimethylarsenate
Sodium dodecylbenzenesulfonate
Sodium fluorophosphate
Sodium isopropylxanthate
Sodium metabisulfite
Sodium metavanadate
Sodium methanearsonate
Sodium molybdate
Sodium naphthalenesulfonate
Sodium orthosilicate
Sodium orthovanadate
Sodium oxalate
Sodium palmitate
Sodium pentafluorostannate
Sodium permanganate
Sodium phenolate (Sodium phenate)
Sodium plumbate
Sodium pyrophosphate
Sodium selenate
Sodium selenite
Sodium selenite pentahydrate
Sodium sesquicarbonate (Trona)
Sodium sesquisilicate
Sodium stannate

Sodium tellurite

Sponge, iron dust

Stannic bromide

Stannic chloride, anhydrous

Stannic chloride, pentahydrate

Stannic chromate

Stannic phosphide(s)

Stannochlor*

Stannous acetate

Stannous bromide

Stannous chloride

Stannous chromate

Stannous-2-ethylhexoate

Stannous oleate

Stannous oxalate

Stannous sulfate (Tin sulfate)

Stannous sulfide

Stannous tartrate

Stibene (Antimony hydride)

Strontium arsenate

Strontium arsenite

Strontium bromide

Strontium chromate

Strontium iodide

Strontium monosulfide

Strontium nitrate

Strontium oxalate

Strontium perchlorate

Strontium peroxide (Strontium dioxide)

Strontium phosphide

Strontium tetrasulfide

Subtilisins (Prills, Proteolytic enzyme)

Tantalum oxide

TCHH (TCHTH, Plictran*, Cyhexatin)

TEL

Telluric acid

Tellurium (Metal)

Tellurium compound, n.o.s.

Tellurium dibromide

Tellurium dichloride

Tellurium dioxide

Tellurium disulfide

Tellurium hexafluoride

Tellurium tetrabromide

Tellurous acid

Tersan* 75

Tetrabutyltin

Tetraethyl lead (TEL)

Tetraethyl silicate

Tetraethyltin

Tetramethyl lead

Tetramethylsilane

Tetraphenyltin

Tetrapropyl orthotitanate

Thallium acetate

Thallium chlorate

Thallium compound, n.o.s.

Thallium monoxide

Thallium nitride

Thallium oxide

Thallium sulfate (Thallium sulphate)

Thallium sulfide

Thimerosal*

Thorium dioxide

Tillex*

Tin compounds, inorganic, n.o.s. (Except stannic oxide)

Tin compounds, organic, n.o.s.

Tin tetrachloride (Tin chloride, Stannic chloride)

Tin tetrachloride, pentahydrate

Titanium dioxide, dust

Titanium hydride

Titanium sesquisulfide

Titanium sulfate (Titanium sulphate)

Titanium sulfide

Titanium tetrachloride

TML

Tremolite (Dust or fine particle; Carcinogen)

Tributyl phosphate (TBP)

Tributyltin acetate

Tributyltin chloride

Tricyclohexyltin hydroxide

Triethanolamine methanearsonate

Triethyl antimony

Triethyl arsine

Triethyl bismuthine

Triethylsilanol

Triethyl stibine

Trilead dinitride

Trimercury dinitride

Trimethyl antimony

Trimethyl arsine
Trimethyl bismuthine
Tri-2-methylpentylaluminum
Trimethylstibine
Triphenylstibine
Triphenyltin acetate
Triphenyltin chloride
Triphenyltin hydroxide
Tri-n-propylaluminum
Tripropyl stibine
Trisilane
Trisilyl arsine
Trithorium tetranitride
Trivinyl stibine
Tryparsamide
Tungstate white
Tungsten carbide
Tungsten compounds, n.o.s.
Tungsten diselenide
Tungsten hexacarbonyl
Tungsten hexachloride
Tungsten oxychloride
Tungsten silicate
Tungstic acid
Tungstic oxide
Ultra-Clor*
Uranium dicarbide
Uranium dioxide
Uranyl acetate
Uranyl nitrate

Uranyl nitrate hexahydrate

Vanadium compound, n.o.s.

Vanadium dichloride

Vanadium hexacarbonyl

Vanadium hexacarbonyl, sodium salt

Vanadium oxytrichloride

Vanadium pentoxide (Vanadic acid anhydride)

Vanadium tetrachloride

Vanadium trichloride

Vanadium tetraoxide

Vanadium trioxide

Vanadous chloride

Vanadyl chloride (Vanadium oxydichloride)

Vanadyl sulfate (Vanadium sulphate)

Vinyl trichlorosilane

White arsenic

White lead

Whitherite

Xenon compounds (Oxides, Oxyfluorides, etc.)

Yellow salt

Yttrium arsenide

Zinc acetylide

Zinc ammonium chloride

Zinc ammonium nitrite

Zinc antimonide

Zinc arsenate

Zinc arsenite (ZMA)

Zinc bisulfite (Zinc bisulphite)

Zinc borate

Zinc bromide

Zinc cadmium sulfide

Zinc carbonate

Zinc chloride

Zinc chloride chromated

Zinc chromate (Human carcinogen)

Zinc cyanide

Zinc dichromate

Zinc diethyldithiocarbamate

Zinc dimethyldithiocarbamate

Zinc dioxide

Zinc dithionite (Zinc hydrosulfite)

Zinc dithionite, aqueous (Zinc hydrosulfite, aqueous)

Zinc ethyl (Diethylzinc; Pyrophoric)

Zinc fluoride

Zinc fluorarsenate

Zinc fluoroborate

Zinc fluorosilicate

Zinc formaldehyde sulfoxylate

Zinc formate

Zinc hydrosulfite (Zinc hydrosulphite)

Zinc-2-mercaptobenzothiazole

Zinc naphthenate

Zinc nitrate

Zinc oxide (Fume, dust)

Zinc perborate

Zinc permanganate

Zinc peroxide

Zinc phenate

Zinc-1,4-phenolsulfonate (Zinc phenolsulphonate)

Zinc phosphide

Zinc propionate

Zinc-1,2-propylene bisdithiocarbamate

Zinc resinate

Zinc salts of dimethyldithiocarbamic acid

Zinc selenate

Zinc selenide

Zinc selenite

Zinc silicofluoride

Zinc sulfate (Zinc sulphate)

Zinc sulfide

Zinc telluride

Zinc trichlorophenate

Zinc yellow (Buttercup yellow, Zinc potassium chromate)

Zineb*

Zipar*

Ziram*

Zirconium acetate

Zirconium ammonium fluoride

Zirconium boride

Zirconium carbide (Dust or powder)

Zirconium carbonate

Zirconium chloride

Zirconium disilicide

Zirconium glyconate

Zirconium hydride

Zirconium hydroxide

Zirconium lactate

Zirconium naphthenate

Zirconium nitride

Zirconium oxide

Zirconium phosphate

Zirconium picramate

Zirconium potassium fluoride

Zirconium potassium sulfate

Zirconium pyrophosphate

Zirconium silicide

Zirconium sulfate (Zirconium sulphate)

Zirconium sulfide

Zirconium tetraacetylacetonate

Zirconium tetrachloride

Zirconium tetrafluoride

ZMA

RGN 25 NITRIDES (REACTIVE)

Soluble nitrides are caustic and toxic. A few nitrides tend to be water reactive and/or pyrophoric (air reactive). Most nitrides react slowly with water or moist air. Also, these compounds tend to be explosive under a variety of conditions. The nitrides decompose in fire or under heat producing ammonia and basic conditions. Nitrides may be ignited by heat, sparks, or flames and tend to reignite after the fire is extinguished. Fire will produce irritating, corrosive, and/or toxic gases which may cause severe injury or death. Some nitrides, such as silver nitride, are shock and vibration sensitive. Avoid breathing vapors from the chemicals or any decomposition products of these compounds. Experienced individuals should handle these chemicals properly. Stable nitrides, which are usually water insoluble, are included in Group (RGN) 24.

Antimony nitride

Cadmium nitride (Tricadmium dinitride)

Calcium nitride (Ticalcium dinitride)

Copper nitride

Lead nitride

Lithium nitride

Mercury nitride

Nitrides, n.o.s.

Potassium nitride

Silver nitride
Sodium nitride
Tetrasulfur tetranitride
Thallium nitride
Tricesium nitride
Trilead dinitride
Trimercury dinitride
Trithorium tetranitride

RGN 26 NITRILES, CYANIDE-FORMING ORGANICS, AND OTHER CYANIDE-CONTAINING ORGANICS

Nitriles are flammable, toxic, and corrosive. Most of these compounds form toxic gases such as hydrogen cyanide and other nitrogen containing substances during high temperature decomposition and/or during reaction with acids. In excess air (oxygen) and/or under basic conditions, the cyanide decomposition products in the presence of heat will often oxidize to a less toxic cyanate or carboxylate species and consequently be much less dangerous. In deficient oxygen and/or under acidic conditions, toxic hydrogen cyanide is favored as a product of combustion and decomposition. During a fire in an oxygen-deficient atmosphere, vapors are usually fatal if inhaled, ingested, or absorbed through the skin. Vapors may form explosive mixtures with air and travel to a source of ignition and flashback. A few of these compounds tend to explode in nitric acid. Avoid breathing vapors from the chemicals or any decomposition products of these compounds.

Acetaldehyde cyanohydrin

Acetamiprid

Acetone cyanohydrin

Acetonitrile

Acrylonitrile

Adiponitrile

Allyl cyanide

Aminopropionitrile

Amyl cyanide

Axall*

Azobisisobutyronitrile

2,2'-Azodi-(2,4-dimethyl-4-methoxyvaleronitrile)

2,2'-Azodi-(2,4-dimethylvaleronitrile)

1,1'-Azodi-(hexahydrobenzonitrile

Azodiisobutyronitrile

2,2'-Azodi-(2-methylbutronitrile)

Baythroid

Benzaldehyde cyanohydrin

Benzonitrile

Benzyl cyanide

Beret*

Brittox*

Bromoacetone cyanohydrin

Bromobenzyl cyanide(s)

Bromoxynil

Bulldock*

Butanenitrile

3-Butenenitrile

Butyronitrile

Carbylamine

CECA

Celest*

Chinmix*

Chlorfenapyr

Chloroacetonitrile

Chloroacrylonitrile

Chlorobenzyl cyanide

Chlorobenzylidene malononitrile (CS, OCBM)

Chlorobutyronitrile

Chloropropionitrile

Chlorothalonil

Chloroxynil

CS gas (Aerosol)

Cyanazine

Cyanoacetamide

Cyanoacetic acid

Cyanochloropentane

Cyanoethyl acrylate

Cyanofenphos

Cyanogas*

Cyanogen, gas or liquid

Cyanomethyl acetate

Cyano(methylmercuri)guanidine

(S)-Cyano(3-phenoxyphenyl)methyl-(S)-chloro-*alpha*-(1-methylethyl)benzeneacetate

Cyanophenphos

Cyanophos (Ciafos)

Cyanopyridine

Cyanthoate

Cyclosal* (Cycloprothrin)

Cyfluthrin

Cyhalotrin

Cymoxanil

Cypendazole

Cypermethrin

Cyphenothrin (Cyphénothrine)

Decamethrin

Deltamethrin

Diallyl cyanamide

Diallylmelamine

3,5-Dibromo-4-hydroxybenzonitrile

Dichlobenil

Dichlorobenzonitrile

Dichloro-5,6-dicyanobenzoquinone (DDQ)

Digital*

Dimethylaminoacetonitrile

Dimethyl cyanamide

Diphenatrile

Diphenylacetonitrile

Diphenylcyanoarsine

Eleeter*

Epoxycyclohexane carbonitrile

Esfenvalerate

Ethoxymethylenemalononitrile

Ethoxyquin

Ethyl cyanide

Ethyl cyanoacetate

Ethyl cyanoacrylate

Ethyl N,N-dimethylphosphoramidocyanidate

Ethylene cyanide

Ethylene cyanohydrin

Ethylenediaminetetraacetonitrile (EDTAN)

Ethylhexyl cyanoacetate

Fenbuconazole

Fenpiclonil

Fenpropathrin

Fenvalerate

Flucythrinate
Fludioxonil
Fluvalinate
Formonitrile
Full*
Fury*
Gambit*
Glutaronitrile
Glycolonitrile
Gokilaht*
Hydrogen cyanamide
Hydroxyisobutyronitrile
Hydroxypropionitrile
Ioxynil
IPM
Isobutyronitrile
Isonitrile
Isopropoxypropionitrile
Isopropyl cyanide
Lactonitrile
Lithic acid (Uric acid)
Lymoxanil
Malonic dinitrile
Malononitrile
Mandelonitrile (Laetrile*)
Maxim*
Maytril*
Medallion*
Methacrylonitrile
Methylacrylonitrile

Methyl cyanide

Methyl cyanoacetate

Methyl-2-cyanoacrylate

Methyl cyanoformate

Mospilan*

Myclobutanil

Neostomosan*

Nitriles, n.o.s.

Nitriles and Organic Cyanides, n.o.s.

Nitrobenzyl cyanide

ODPN (Oxydipropionitrile)

Phenylacetonitrile

Phenyl valerylnitrile

Phoxim

m-Phthalodinitrile

Phthalonitrile

Pirate*

Polyacrylonitrile decomposition products (Orlon, Dynel, etc.; High temperature stability of polyacrylonitriles)

Propanenitrile

Propenenitrile

Propionamide nitrile

Propionitrile (Ethyl cyanide)

Propyl cyanide (Butyronitrile)

Responsar*

Ricinine (Ricin)

SAGA*

Santoquin*

Saphire*

Scout X-TRA

Surecide*

Tabun

Tetrachloroisophthalonitrile

Tetracyanomethylene

Tetramethyl succinonitrile (TMSM)

Thalonex*

Thiocarboxime

β,β-Thiopropionitrile

Tralex*

Tralomethrin (Scout-X-TRA*)

Transmix*

Trichloroacetonitrile

2,6,8-Trioxypurine

Tritox*

Uric acid (Uric oxide)

Vinyl cyanide

Zeta-Cypermethrin

RGN 27 NITRO COMPOUNDS (ALL ISOMERS)

These compounds are highly flammable, while a few compounds are combustible. They are easily ignited by heat, sparks, or flames and tend to form explosive mixtures in air. Vapors may travel to the source of ignition and flash back. The compounds are a vapor explosion hazard indoors, outdoors, and in confined spaces and sewers. Some of these compounds may explode from friction, shock, heat, and acid contamination. Some compounds will ignite or react with wood, paper, oil, clothing, fuels, hydrocarbons, etc. A number of these compounds will form dangerous nitrogen oxide and/or toxic acid gases during high temperature decomposition. A few compounds tend to decompose slowly at room temperature, whereas other chemicals require fire or much higher temperatures before decomposition begins. Several compounds are less toxic than the decomposition products formed during a fire or catastrophic event. Avoid skin contact and breathing vapors from the chemicals or any decomposition products of these chemicals. Many of these organic compounds will oxidize other materials. Experienced individuals should handle these chemicals properly.

Acetyl nitrate

2-Amino-4,6-dinitrophenol

Ammonium dinitro-*o*-cresolate

Amyl nitrate

Bayluscid

Benefin

Binapacryl

Bromethalin

2-Bromo-2-nitropropane-1,3-diol

Bromopicrin

Bronopal

Brotal*

Bulan

Butamifos

Butralin

2-(*sec*-Butyl)-4,6-dinitrophenol

Butyl nitrate (*tert*-, *iso*- *n*-)

Butyl nitrite

5-*tert*-Butyl-2,4,6-trinitro-*m*-xylene

Cellulose nitrate

Chlomethoxyfen

Chlorodinitrobenzene

Chlorodinitrotoluene

Chloro-*m*-nitroacetophenone

Chloronitroaniline(s)

Chloronitrobenzene

Chloronitrobenzenesulfonic acid

Chloronitrobenzoic acid

Chloronitrobenzotrifluoride

Chloronitropropane (Korax)

Chloronitrotoluene(s)

Chloropicrin (Chlorpicrin)

Chlorotrinitrobenzene

Cobra*

Collodion

Cremart*

Cyclonite

Cyclotrimethylenetrinitramine

DCNA

DDNP

DEGN

Diazodinitrophonol

Dicapthon

Dichloronitrobenzene

1,1-Dichloro-1-nitroethane

2,4-Dichlorophenyl-4-nitrophenyl ether

Dicloran

Diethylene glycol dinitrate (DEGN)

Dilan

Dimethylnitrobenzene

O,O-dimethyl-*O*-*p*-nitrophenyl phosphorothioate

Dinitolmide (3,5-Dinitro-*o*-toluamide, Zoalene)

Dinitramine

Dinitroaminophenol

Dinitroaniline(s)

Dinitroanisole

Dinitrobenzene(s)

Dinitrobenzoyl chloride

2,4-Dinitro-6-*sec*-butylphenol (DNBP, Dinoseb)

Dinitrochlorobenzene (Dinitrochlorbenzene)

Dinitrocresol

Dinitrocyclohexylphenol

Dinitrofluorobenzene (DNFB)

Dinitronaphthalene

Dinitro-*o*-*sec*-amylphenol (DNAP)

Dinitrophenol

Dinitrophenolate compounds
Dinitrophenylhydrazine
1,6-Dinitropyrene
1,8-Dinitropyrene
Dinitroresorcinol
Dinitrosalicylic acid
3,5-Dinitro-*o*-toluamide
Dinitrotoluene(s)
2,4-Dinitrotoluene (DNT)
Dinobuton
Dinocap
Dinocton
Dinopenton
Dinoprop
Dinosam (Dinosame)
Dinoseb
Dinoseb acetate
Dinoseb-ammonium
Dinosulfon
Dinoterb
Dinoterb acetate
Dipentaerythritol hexanitrate
Dipicrylamine
Dipicryl sulfide
Dipropalin
DNAP
DNOC
DPC
EPN
Ethalfluralin

Ethylene glycol dinitrate (Low freezing)

Ethyl nitrate

Ethyl nitrite

Ethyl parathion

Etinofen

Faneron*

Fluazinam

Flumetralin

Fluorodifen

Fomesafen

Furazolidone

Glycol dinitrate

Glycol monolactate trinitrate

Guanidine nitrate

Gun cotton (Nitrocellulose)

Hexanitrodiphenylamine (Hexite, Hexil, Dipicrylamine)

Hexanitrodiphenyl sulfide

4-Hydroxy-3-nitrobenzenearsonic acid

Imidacloprid

Isosorbide dinitrate (Mixture)

Isosorbide-5-mononitrate

Korax

Lacquer

Lactofen

Lead dinitroresorcinate

Lead mononitroresorcinate

Lead styphnate

Lead trinitroresorcinate

Mannitol hexanitrate (HNM)

Medinoterb acetate

Metafos

N-Methyl-*N'*-nitro-*N*-nitrosoguanidine

Methyl parathion

Mononitrotoluidine(s)

Naphite

Nichlorfos

NiPar S-10* and NiPar S-20*

Nitranilic acid

Nitroaniline

Nitrobromobenzene

Nitrobromoform

Nitrocellulose (Wet or dry)

Nitrochlorobenzene

3-Nitro-4-chlorobenzotrifluoride

Nitrochloroform

Nitrodiphenyl (Nitrobiphenyl; Various forms)

Nitroethane

Nitrofen

Nitrogil*

Nitroglycerin

Nitroguanidine

Nitromannite

Nitromethane

Nitrophenol

Nitropropane (Various forms)

2-Nitropropane

1-Nitropyrene

4-Nitropyrene

Nitrostarch, dry

Nitrostyrene

Nitrotrichloromethane
Nitrourea
Nitroxylol
Organic pigment(s), self-heating
Oryzalin
Oxyfluorfen
Paraoxen (Para-oxen)
Parathion
PCNB
Pendimethalin
Pentachloronitrobenzene (PCNB)
Pentaerythritol tetranitrate (PETN)
Pentolite
PETN
Phenol trinitrate
Phosnichlor
Picramic acid (Picraminic acid)
Picramide
Picric acid (Picronitric acid)
Picrite, wetted
Picryl chloride
Polyvinyl nitrate
Potassium dinitrobenzfuroxan
Prodiamine
Prolan
Propylene glycol dinitrate (PGDN)
Pyroxylin
Quintozene
RDX
2-(Sec-butyl)-4,6-dinitrophenol

Silver styphnate
Silver trinitroresorcinate
Sodium dinitro-o-cresylate
Sodium picramate
Sonalan*
Starch nitrate
Styphnic acid
Substituted nitrophenol pesticides
Surflan*
TCNA
Team*
Terrazan*
Terr-O-Gas*
Tetrachlorodinitroethane
Tetrachloronitroanisole
Tetranitroaniline (TNA)
Tetranitromethane
Tetryl
Thiophos (Parathion)
TNA
TNB
TNT
Tok*
Treflan*
Trialin*
Trichloronitromethane
Trifluralin
Trinitroanisole
Trinitrobenzene (TNB)
Trinitrobenzoic acid

2,4,6-Trinitro-*m*-cresol

Trinitroglycerin

Trinitromethane (In air)

Trinitronaphthalene

Trinitrophenol

Trinitrophenyl methyl ether

Trinitrophenylmethylnitramine

Trinitroresorcinol

2,4,6-Trinitrotoluene (TNT)

Tri-Scept*

Urea nitrate

Zoalene

Zoamix*

RGN 28 HYDROCARBONS, ALIPHATIC, UNSATURATED (ALL ISOMERS)

Hydrocarbons are usually highly flammable, but the higher molecular weight hydrocarbons can be combustible. The compounds can be easily ignited by heat, sparks, or flames. Some vapors form explosive mixtures with air. Vapors may travel to the source of ignition and flashback. These compounds are an explosion hazard indoors, outdoors, and in confined spaces and sewers. Containers can explode when heated. Some compounds may polymerize explosively when heated or during a fire. Some of the alkyne-type compounds, such as butynediol and acetylene, tend to explode with contamination by mercury and silver salts, strong acids, etc.

Acetylene, gas

Allene, gas or liquid

Amylene

Bicyclo[2.2.1]hepta-2,5-diene

Bromoacetylene

Butadiene, gas or liquid

Butadiyne, gas or liquid

Butene, gas

Butylene, gas

Butyne, gas or liquid

1,4-Butynediol

Chlorobutadiene

Chloroprene (Gas or Liquid)

1-Chloropropene and 3-Chloropropene

2-Chloropropene, gas or liquid

Chloropropyne

Croton oil

Crotonylene

Crotyl alcohol

Cyclobutene

Cyclododecatriene

Cycloheptatriene

Cycloheptene

Cyclohexadiene

Cyclohexene

Cyclohexene oxide

Cyclooctadiene(s)

Cyclooctatetraene

Cyclopentadiene

Cyclopentene

Decene

DES

Diacetylene, gas or liquid

Dichloroacetylene

Dichlorobutene

Dichlorodifluoroethylene

Dichloroethylene

Dichloropropene (Dichloropropylene)

1,3-Dichloropropene

Dichloropyrene(s)

Dicycloheptadiene

Dicyclohexadiene

Dicyclopentadiene

Diethylstilbestrol

Diiodacetylene (Diiodethyne)

Diiododibromoethylene

Diisobutylene

Dimethylacetylene

Dimethylbutene

Dimethylbutyne (All forms)

2,5-Dimethyl-2,5-di(*tert*-butylperoxy)hexyne-3

Dimethylheptene

Dimethylhexadiene

Dimethylhexynol

Dimethyloctynediol

Dipentene

Dipropargyl

Dodecene

Ethchlorvynol

Ethylacetylene

Ethylene (Ethene)

Ethylhexene (Various forms)

5-Ethylidene-2-norbornene (ENB)

Ethyl propiolate

Ethyne

Heptene(s)

Hexachlorobutadiene

Hexachlorocyclopentadiene

Hexadiene(s)

Hexadiyne

Hexene(s)

Hexyne(s)

Hexynol

Hydrocarbon liquid, alkene or alkyne, n.o.s.

Hydrocarbons, aliphatic, n.o.s.

Imiprothrin

Isobutene, gas or liquid (Isobutylene)

Isoheptene

Isohexene

Isooctene

Isopentene

Isoprene

Isopropenylacetylene

Isopropyl acetylene

Limonene

Methylacetylene

Methylacetylene-propadiene, stabilized (MAPP)

Methylbutadiene

Methylbutene

Methyl butyne

Methylcyclohexene (Various forms)

Methylpentadiene

Methylpentene (Various forms)

Methyl styrene

Nonene

2,5-Norbornadiene

Ocadecyne

Octadiene

Octafluorobut-2-ene

Octene

Pentadiene

Pentene
Pentyne
alpha-Phellandrene
Phenylethylene
(*alpha*-, *beta*-) Pinene
Piperylene
Polybutadiene
Polybutene
Polypropylene
Polystyrene
Propadiene
Propargyl alcohol
Propargyl bromide
Propargyl chloride
Propene
Propylene
Propylene tetramer
Propyne
Pyrene
Quinine
Safrole
Solvenol*
Styrene, monomer
Terbene
Terpene
Terpinolene
Tetradecene
Tetrahydronaphthalene
Triallyl cyanurate
Trichlorobutene

Tridecene

Triisobutylene

Trimethylpentene

Tripropylene

Undecene

Vinylacetylene

Vinylbenzene

Vinylcyclohexene

Vinylcyclohexene dioxide

Vinylcyclohexene monoxide

Vinyltoluene

RGN 29 ALKANES OR HYDROCARBONS, ALIPHATIC, SATURATED

Aliphatic hydrocarbons or alkanes are usually highly flammable, but sometimes combustible. The compounds can be easily ignited by heat, sparks, or flames. Some vapors form explosive mixtures with air. Vapors may travel to the source of ignition and flashback. They are an explosion hazard indoors, outdoors, and in confined spaces and sewers. Containers can explode when heated. In a fire, some compounds react vigorously with oxidizing materials. A number of these compounds can cause respiratory paralysis and death. Avoid inhalation of vapors and prolonged skin contact. In a few cases, stable ethers are similar to alkanes.

Butane

Cyclobutane

Cycloheptane

Cyclohexane

Cycloparaffin

Cyclopentane

Cyclopropane

Decalin

Decane

Dimethylbutane

Dimethylcyclohexane

Dimethylpentane

2,2-Dimethylpropane

Ethane

Ethylcyclohexane

Ethylcyclopentane

Heptane

Hexane

Hydrocarbon, aliphatic, n.o.s.

Isobutane

Isoheptane

Isohexane

Isooctane

Isopentane

Methane

Methylbutane

Methylcyclohexane

Methylcyclopentane

Methylheptane

Methylhexane

Methylpentane

Methyltetrahydrofuran

Neohexane

Neopentane

Nitroethane

Nitroparaffin

Nonane(s)

Octane(s)

Pentamethylheptane

Pentane(s)

Propane

Silyl compounds, n.o.s.
Tetradecane
Trimethylbutane
Trimethylhexane
Trimethylpentane
Undecane
Vinylcyclohexane

RGN 30 ORGANIC PEROXIDES, HYDROPEROXIDES, AND ORGANIC OXIDIZING AGENTS

These toxic compounds are a fire risk and an explosion hazard. These compounds may be ignited by heat, sparks, and flame. Many of the organic peroxides are sensitive to heat, shock, friction, and contamination. Several organic peroxides explode and/or decompose easily, while other organic peroxides ignite spontaneously in air. A number of these compounds are an extreme explosion hazard at relatively low temperatures. Some peroxides ignite wood, paper, oil, cloth, etc. These compounds are strong oxidizing agents and should not be stored near combustible materials or reducing agents. Avoid breathing vapors from these compounds or decomposition products of the chemicals. Experienced individuals should handle these chemicals properly.

Acetyl acetone peroxide

Acetylbenzoyl peroxide

Acetyl cyclohexanesulfonyl peroxide

Acetyl peroxide

t-Amyl peroxy-2-ethylhexanoate

t-Amyl peroxyneodecanoate

Benzoyl peroxide

Bis(1-hydroxycyclohexyl)peroxide

N-Bromoacetamide (NBA)

n-Butyl-4,4-di-(*tert*-butylperoxy)valerate

t-Butyl cumene peroxide

t-Butyl cumyl peroxide

Butyl hydroperoxide (*tert*-, *iso*-, *n*-)

tert-Butylisopropylbenzene hydroperoxide

t-Butyl monoperoxymaleate

t-Butyl perbenzoate

Butyl peroxide (*tert*-, *iso*-, *n*-)

Butyl peroxyacetate

Butyl peroxybenzoate

Butyl peroxycrotonate

Butyl peroxydicarbonate

Butyl peroxydiethylacetate

Butyl peroxy-2-ethylhexanoate

Butyl peroxyisobutyrate

Butyl peroxyisononanoate

Butylperoxyisopropyl carbonate

Butylperoxymaleic acid

Butyl peroxyneodecanoate

Butyl peroxy-3-phenylphthalide

Butylperoxyphthalic acid

Butylperoxypivalate

Butyl peroxy-3,5,5-trimethylhexanoate

Caprylyl peroxide

Carbamide peroxide

Chlorobenzoyl peroxide

Chloroperoxybenzoic acid

Cumene hydroperoxide

Cyclohexanone peroxide

Decanoyl peroxide

Diacetone alchol peroxide

Dibenzoyl peroxide

Dibenzyl peroxydicarbonate

Di-(4-*tert*-butylcyclohexyl)peroxydicarbonate

Di-*tert*-butyl diperphthalate

Di-*tert*-butyl peroxide (DTBP)

2,2-Di-(*tert*-butylperoxy)butane

1,1-Di-(*tert*-butylperoxy)cyclohexane

Di-(*sec*-butyl)peroxydicarbonate

1,3-Di-(2-*tert*-butylperoxyisopropyl)benzene

1,4-Di-(2-*tert*-butylperoxyisopropyl)benzene

Di-(*tert*-butylperoxy)phthalate

2,2-Di-(*tert*-butylperoxy)propane

1,1-Di-(*tert*-butylperoxy)-3,3,5-trimethyl cyclohexane

Dicetyl peroxydicarbonate

2,4-Dichlorobenzoyl peroxide

Dicumyl peroxide

2,2'-Di-(4,4-di-*tert*-butylperoxycyclohexyl)propane

Di-(2-ethylhexyl)peroxydicarbonate

Diethyl peroxydicarbonate

2,2-Dihydroperoxypropane

Di-(hydroxycyclohexyl)peroxide

Diisobutyryl peroxide

Diisopropylbenzene hydroperoxide

Diisopropyl peroxydicarbonate

Diisotridecyl peroxydicarbonate

Di-(2-methylbenzoyl)peroxide

Dimethylbenzyl hydroperoxide

2,5-Dimethyl-2,5-di(benzoylperoxy)hexane

2,5-Dimethyl-2,5-di(*tert*-butylperoxy)hexane

2,5-Dimethyl-2,5-di(*tert*-butylperoxy)hexyne-3

2,5-Dimethyl-2,5-di-(2-ethylhexanoylperoxy)hexane

2,5-Dimethylhexane-2,5- dihydroperoxide (Dimethylhexane dihydroperoxide)

2,5-Dimethylhexane-2,5-diperoxybenzoate (Dimethylhexane diperoxybenzoate)

Dimyristyl peroxydicarbonate

Di-*n*-propyl peroxydicarbonate

Distearyl peroxydicarbonate

Di-(3,5,5-trimethyl-1,2-dioxolanyl-3)peroxide

Ethyl-3,3-di-(*tert*-butylperoxy)butyrate

3,3,6,6,9,9-Hexamethyl-1,2,4,5-tetraoxacyclononane

Hydroperoxide, organic, n.o.s.

IPP (Isopropyl percarbonate)

Isononanoyl peroxide

Isopropyl percarbonate

Isopropyl peroxydicarbonate

Lauroyl peroxide

p-Menthane hydroperoxide

Methyl ethyl ketone peroxide

Methyl isobutyl ketone peroxide

Myristoyl peroxide

Octanoyl peroxide

Octyl peroxide

Organic peroxide, n.o.s.

Pelargonyl peroxide

Peracetic acid

Peroxides, organic, n.o.s.

Peroxyacetic acid

Pinane hydroperoxide

Propionyl peroxide
Succinic acid peroxide
TATP
Tetralin hydroperoxide
1,1,3,3-Tetramethylbutyl hydroperoxide
1,1,3,3-Tetramethylbutyl peroxy-2-ethylhexanoate
Urea peroxide

RGN 31 PHENOLS, CRESOLS, CATECHOLS, RESORCINOL, PHENOXY-TYPE PESTICIDES, AND STABLE AROMATIC ETHERS (ALL ISOMERS).

These chemicals are flammable or combustible. A number of these chemicals containing nitro-, chloro-, bromo-, or sulfur groups can produce toxic and/or acidic gases during a fire. Avoid breathing vapors from these chemicals or decomposition products of the chemicals.

para-Acetylaminophenol

Alkyl phenols, n.o.s.

Aminophenol(s)

Benzyl chlorophenol

Bromol

Bromophenol

Bromophenylphenol

para-tert-Butylcatechol

tert-Butyl-*meta*-cresol (MBMC)

2-(*sec*-Butyl)-4,6-dinitrophenol

Butylphenol(s)

Butylphenyl ether

tert-Butyl-*ortho*-thiocresol

4-*tert*-Butylthiophenol

Carbolic acid (Carbolic oil)

Carvacrol

Catechol

Chloro-*tert*-amylphenol

Chlorocresol(s)

Chloromethylphenol

Chlorophenolate(s)

Chlorophenol(s)

Chlorophenylphenol(s)

Chlorothymol

para-Chloro-*meta*-xylenol

Chloroxylenol(s)

Coal tar pitch volatile(s)

Cresols, n.o.s.

Creosote (Coal, wood)

Cresylic acid

Cyclohexylphenol

Diaminophenol

Diamyl phenol

Di-*tert*-butyl-*para*-cresol

Di-*tert*-butylphenol

Dichlorophenol

Diethylaminophenol

Dihydroxybenzene

Dimethylaminomethyl phenol

Dimethylphenol

2,4-Dinitro-6-*sec*-butylphenol

Dinitrocresol

Dinitrocyclohexylphenol

Dinitro-*o-sec*-amylphenol (DNAP)

Dinitrophenol

Dinitroresorcinol

Dinonylphenol

Dinoprop

Dinosam

Dinoseb

Dinoterb

DNAP

DNOC

Dodecylphenol

Ethylphenol

Etinofen

Exxpel 2*

Fluorophenol

Gallic acid

Gossypol (Polyphenol)

Guaiacol (Methyl catechol)

2-HAP

Hexylresorcinol

Hydroquinone

Hydroquinone monomethyl ether

Hydroxyacetophenone

Hydroxydiphenol

Hydroxyhydroquinone

4-Hydroxymethyl-2,6-di-*tert*-butylphenol

Hydroxyphenol(s)

Isoeugenol

Isopropylcresol

Isopropylphenol

MBMC

Methoxyphenol

Naphthol

Nitrocresol

Nitrophenol

PCP

Pentachlorophenol

Phenol

Phenolates, n.o.s.

Phenolsulfonic acid

Phenol trinitrate

Phenoxyacetic acid derivative pesticide

Phenoxy pesticide

2-Phenyl-6-chlorophenol

Phenylphenol (Various forms)

Phloroglucinol

Picric acid

Picronitric acid

Picrite, wetted

Pyrogallol

Resorcinol

2-(*sec*-Butyl)-4,6-dinitrophenol

Septiphene

Sodium phenolsulfonate

Substituted nitrophenol pesticides

2,4,6-T

Tar acid

Terbec*

Tetrabromo-*o*-cresol

Tetrachlorophenol

4,4'-Thiobis(6-*tert*-butyl-*m*-cresol)

Thiodiphenol (TDP)
Tribromophenol
Trichlorophenol
2,4,6-Trichlorophenol
Trihydroxybenzene
Trihydroxybenzoic acid
2,4,6-Trinitro-*m*-cresol
Trinitrophenol
Urushiol
Xylenol(s)

RGN 32 ORGANOPHOSPHATES, PHOSPHOTHIATES, PHOSPHAMIDES, AND PHOSPHODITHIOATES

These compounds contain a number of different types of species (chlorine, bromine, sulfur, nitrogen, phosphorus) within the chemical structure. Highly toxic substances can be inhaled or absorbed through the skin. Many chemicals are *cholinesterase inhibitors* and/or cause nerve damage in humans. During a fire, the potential for formation of a variety of toxic gaseous substances is probable. Fire will produce irritating, corrosive, and/or toxic gases. Avoid breathing vapors from these chemicals or from decomposition products of the chemicals. Some of the vapors can be highly flammable and can be easily ignited by heat, sparks, or flames. The vapors can travel to a source of ignition and flashback. Vapor explosions and poison hazards are potential problems. Also, containers may explode when heated.

Abate* (Abathion)

Acephate

Acetoxon (Acethion, Azethion)

Acifat*

Acifon*

Afugan*

Aimcosystox*

Aminophos

Anilofos

Anthio*

Apache*

Azidithion

Azinphos ethyl

Azinphos methyl

1-Aziridinyl phosphine oxide (Tris)

1-Aziridinyl phosphine sulfide (Tris)

Azition

Azodrin*

Baytex*

Bensulide

BFPO

Bidrin*

Bladafum*

Bolstar*

Bomyl*

Bromchlophos

Bromophos

Butamifos

Butonate

Cadusafos

Camphechlor

Carbofos

Carbophenothion

Chinalphos

Chlorethoxyfos

Chlorfenvinphos

Chlormephos

Chloroethoxyfos

Chlorofenvinphos

Chlorofos* (Chlorophos)
Chlorpyrifos (Chlorpyriphos)
Chlorthion
Chlorthiophos
Coroxon
Cotnion-Methyl*
Coumaphos
Cremart*
Crotoxyphos
Cruformate
Crysthyon*
Cyanofenphos (Cyanophenphos)
Cyanophos
Cyanthoate
Cyclophosphamide
Cythion
Cytrolane
DDVP
DEF 6*
Delnav*
Delsam*
Demephion-S (Demephion-O)
Demeton
Demeton-O-methyl
Demeton-S-methyl
2,4-DEP
Dialifor
Diamidafos
Diazinon*
Dicapthon

Dichlofenthion

O-(2,4-Dichlorophenyl)-O,O-diethyl phosphorothioate

o-(2,4-Dichlorophenyl)-o-methylisopropylphosphoramidothioate (DMPA)

Dichlorvos (2,2-Dichlorovinyldimethyl phosphate)

Dicrotophos

Diethon

Diethyl chlorophosphate

Diethyl chlorovinyl phosphate

Diethyl-1-(2,4-dichlorophenyl)-2-chlorovinyl phosphate

O,O-Diethyl phosphorochloridothioate (Ethyl PCT)

Diethylthiophosphoryl chloride

O,O-Diethyl-O-3,5,6-trichloro-2-pyridylphosphorothioate

N,N-Diisopropyldiamidophosphoryl fluoride (Mipafox)

Diisopropyl fluorophosphate

Dimefox

Dimethoate

(Dimethoxyphosphinyloxy)-N,N-dimethyl-cis-crotonamide

Dimethyl chlorothiophosphate

Dimethyl-1,2-dibromo-2,2-dichloroethyl phosphate

Dimethyldichlorovinyl phosphate

Dimethyldithiophosphoric acid

O,O-Dimethyl-S-2-(ethylsulfinyl)ethylphosphorothioate

O,O-Dimethyl-O-p-nitrophenyl phosphorothioate

O,O-Dimethyl phosphorochloridothioate (Methyl PCT)

Dimethylthiophosphoryl chloride

Dimeton

Diothyl

Dioxathion (Delnav*)

Dipterex*

Disulfoton

Di-Syston*

Ditalimfos

Dizalin* (Dizalux*)

Doom*

Dursban*

Dyfonate*

Dylox*

Edifenphos

Endothion

EPN

Estox

Ethion

Ethoprop (Ethoprophos)

O-Ethyl-*S*-(2-diisopropylaminoethyl)methylphosphonothiolate

Ethyl *N,N*-dimethylphosphoramidocyanidate

Ethyl parathion

Ethyl PCT

Ethyl phosphonothioc dichloride, anhydrous

Ethyl phosphonous dichloride, anhydrous

Ethyl phosphorodichloridate

Ethylthiodemeton

Exothion

Famphur

Fenamiphos

Fenchlorphos

Fenitrothion

Fenophosphon

Fensulfothion

Fenthion
Fonofos* (Fonophos)
Formothion
Fortress*
Fosfamid
Fosthiazate
Fostion*
Glyphosate
Glyphosate isopropylamine salt
Glyphosate-trimesium
Gusathion A*
Guthion*
Hanane*
Heptenophos
Hercon* Vaportape II (Dichlorovos)
Hexaethyl tetraphosphate (HETP)
Hexaethyl tetraphosphate, compressed gas mixture
Hexamethylphosphoramide (Hempa)
Hostaquick*
IBP
Iprobenfos
Isazofos
Isofenphos (Isophenphos)
Isopropylamine glyphosate
Isopropyl methylphosphonofluoridate
Isopropyl methylphosphoryl fluoride
Isothioate
Isoxathion
Kayaphos*
Kilvale*

Luxathion*
M-81
Malathion (Malatop*, Maldison)
Mecarbam
Mecarphon
Menazon
Mephosfolan
N-(2-Mercaptoethyl)benzenesulfonanide
Mercaptofostiol
Mercaptothion (Mercaptotion)
Metafos
Metasystox-R
Metasystox-S
Methamidophos
Methidathion
Methyl demeton
Methyl mercaptofos
Methyl parathion
Methyl phencapton
Methyl potasan
Mevinphos
Mipafox
Mocap*
Monacide*
Monocrotophos
Morphothion
Morzid
MSR2*
Naled
Nemacur*

Nerve gases, various types

Nichlorfos

Nogos*

Novachem

Novartis*

Nuvan*

Octamethylpyrophosphoramide

Oftanol*T

Ofunack*

Omethoate

Organic phosphate compound, n.o.s. (Organophosphate compound, n.o.s.)

Organic phosphorus compound, n.o.s. (Organophosphorus compound, n.o.s.)

Organophosphorus pesticide, n.o.s.

Orthene*

Ortran*

Oxydemeton-methyl

Paraoxon

Parathion

Phenthoate

Phorate

Phosalone

Phosdrin*

Phosmet

Phosnichlor

Phosphamidon

Phosphodithioate(s), n.o.s.

Phospholan (Phosfolan)

Phosphothiate(s), n.o.s.

Phoxim (Phoxime)

Phthalofos

Piperophos

Pirimiphos (Pirimiphos-methyl, Pirimiphos-ethyl)

PMP

Polirac*

Polychlorcamphene

Potassan

Pre-San*

Primicid

Profenofos

Propaphos

Propetamphos

Propoxon

Prothidathion

Prothiofos (Prothiophos)

Prothion

Prothoate

Pyraclofos

Pyrazophos

Pyridaphenthion

Pyrimiphos-ethyl (Pyrimiphos-methyl)

Pyrimithate

Pyrinox*

Quinalphos

Rilof*

Rizolex*

Ronnel

Roundup

Rugby*

Sarin (Nerve gas)
Smite*
Soman
Sophamide
Sulfotep (Sulfotepp)
Sulprofos
Supracide*
Systox
Tabun
Tebupirimfos
TEDP
Temephos
Tera (APO, Apoxide)
TEPP
Terbufos
Terraguard*
Tetrachlorovinphos
Tetraethyl dithiopyrophosphate
Tetraethyl pyrophosphate
Tetra-n-propyl dithionopyrophosphate
Thimet
Thiometon
Thionazin
Thiophos
Thiotepa
Timet
Tokuthion*
Tolclofos-methyl
Touchdown*
Toxaphene with methyl parathion

Triamiphos

Triazophos

Triazotion

O,O,O-Tributyl phosphorothioate

S,S,S-Tributyl phosphorotrithioate (DEF)

Tributyl phosphorotrithioite

Trichlorfon (Trichlorphon)

Trichloronate

Tricresyl phosphate (TCP)

Triethylenephosphoramide (APO, Tepa)

Triethylenethiophosphoramide

Triethyl phosphate

O,O,O-Triethyl phosphorothioate

O,O,O-Triisooctyl phosphorothioate

Trimethyl phosphate

Trimethyl phosphorothionate

Trioctyl phosphate

Triorthocresyl phosphate

Triphenyl phosphate

Tris(1-aziridinyl) phosphine oxide

Tris(1-aziridinyl) phosphine sulfide

Ultracide*

Vamidothion

Vapona* (Vaponite*)

Voltage

Zinophos*

RGN 33 SULFIDES, INORGANIC

These compounds are moderately to highly toxic (poisons). Several compounds are a dangerous fire risk near oxidizing materials. Some materials will form explosive mixtures with air. Under acidic conditions, the compounds will produce the toxic and dangerous gas, hydrogen sulfide.

The liquids and vapors in air have a strong disagreeable odor and have narcotic tendencies on humans. Vapors may be ignited by heat, sparks, or flames. Vapors from these compounds travel to the source of ignition and flash back. Containers can explode when heated. Low boiling liquids and gases are flammable, toxic, and tend to explode easily. Avoid breathing toxic sulfide vapors of the compounds or the toxic gases formed during reaction or during decomposition of the chemical. In a fire, a number of these sulfide compounds can act as a reducing agent and impact the toxicity level of other chemicals. Under excess oxidative conditions in a fire, sulfides tend to form toxic gases such as sulfur dioxide and sulfur trioxide. Some solids are insoluble in water, but form hydrogen sulfide under acidic conditions or during a fire.

Aluminum sulfide

Ammonium bisulfide

Ammonium hydrosulfide (Ammonium hydrosulphide)

Ammonium polysulfide

Ammonium sulfide (Ammonium sulphide)

Antimony pentasulfide

Antimony sulfide (Antimony sulphide)

Antimony trisulfide (Antimony trisulphide)
Arsenic disulfide
Arsenic pentasulfide
Arsenic sulfide (Arsenic sulphide)
Arsenic trisulfide (Arsenic trisulphide)
Barium sulfide
Beryllium sulfide
Bismuth sulfide
Bismuth trisulfide
Boron trisulfide
Cadmium sulfide
Calcium sulfide
Carbon bisulfide (Carbon bisulphide)
Carbon disulfide (Carbon disulphide)
Carbonyl sulfide (Carbonyl sulphide), gas
Cerium trisulfide
Cesium sulfide
Chromium sulfide
Chromic sulfide
Copper sulfide (Cupric sulfide)
Cuprous sulfide
Ferric sulfide
Ferrous sulfide
Germanium sulfide
Gold sulfide
Hydrogen sulfide, gas
Lead sulfide
Lithium sulfide
Magnesium sulfide
Manganese sulfide

Mercuric sulfide
Molybdenum sulfide
Nickel subsulfide
Phosphoric sulfide
Phosphorous heptasulfide
Phosphorous pentasulfide
Phosphorous sesquisulfide
Phosphorous trisulfide
Potassium hydrosulfide (Potassium hydrogen sulfide)
Potassium polysulfide
Potassium sulfide, anhydrous
Potassium sulfide (With water)
Red arsenic
Selenium sulfide (Selenium disulfide)
Silver sulfide
Sodium hydrosulfide
Sodium polysulfide
Sodium sulfide
Sodium tetrasulfide
Stannic sulfide
Stannous sulfide
Strontium monosulfide
Strontium sulfide
Strontium tetrasulfide
Sulfides, inorganic/metal, n.o.s.
Tellurium disulfide
Tetraphosphorus trisulfide
Thallium sulfide
Titanium disulfide (Titanium disulphide)
Titanium sesquisulfide

Titanium sulfide
Uranium sulfide
Vanadium sulfide
Zinc cadmium sulfide
Zinc sulfide
Zirconium sulfide

RGN 34 EPOXIDES

Epoxides are highly flammable and reactive. The vapors may be fatal if inhaled or absorbed through skin. Fire will produce corrosive, irritating, and toxic substances. These compounds can form explosive mixtures in air. Vapors can travel to the source of ignition and flash back. Containers can explode when heated. Some of the materials may react violently with water.

Experienced individuals should handle these chemicals properly.

Butyl glycidyl ether

t-Butyl-3-phenyl oxazirane

1-Chloro-2,3-epoxypropane

Chloropropylene oxide

Cresol glycidyl ether

Diepoxybutane

Diglycidyl ether (DGE)

Diglycidyl resorcinol ether

Epibromohydrin

Epichlorohydrin

Epoxy

Epoxybutane

Epoxybutene

Epoxycyclohexane carbonitrile

Epoxyethane

Epoxyethylbenzene

1,2-Epoxy-3-ethoxypropane

Epoxyheptachlor (HCE)

2,3-Epoxypropyl ether or Bis-(2,3-Epoxypropyl) ether

1,2-Epoxypropane

2,3-Epoxy-1-propanol

Ethylene oxide

Glycidol

HCE

Heptachlor epoxide

HFPO

Propylene oxide

Styrene oxide

4-Vinyl-1-cyclohexene diepoxide

RGN 35 FLAMMABLE GASES (DANGEROUS FIRE AND EXPLOSION RISK)

Extremely flammable gases will be ignited by heat, sparks, or flames. The gases will form explosive mixtures with air. The gases will travel to the source of ignition and flash back.

Containers may explode when heated. Some gases are toxic and cause injury, while others are asphyxiants. A number of halogenated (fluorine, chlorine, and bromine) gases form hazardous acid vapors during combustion, which can lead to serious breathing and corrosion problems. Most of these gases do not react with liquids. However, in the presence of a spark, these gases may react with liquids in the vapor phase (within specific ranges of concentration) to produce fire, toxic materials, and explosions.

Acetylene

Allene, gas or liquid

Bromotrifluoroether

Bromotrifluoroethylene

1,3-Butadiene, gas or liquid

Butadiyne

Butane

Butene

Butylene

Butyne

Chlorodifluoroethane(s)

Chloroethylene

Chloromethyl ethyl ether

Chloromethyl methyl ether

Chlorotrifluoroethylene

Cyclobutane

Cyclobutene

Cyclopropane

Deuterium

Diacetylene

Dichlorodifluoroethane (Difluorodichloroethane)

Difluorochloroethane(s) (Difluoromonochloroethane)

Difluoroethane

Difluoroethylene

Difluoromethane

Dimethyl ether, gas or liquid

2,2-Dimethylpropane

Epoxyethane

Ethane

Ethene (Ethylene)

Ethylacetylene

Ethyl chloride

Ethylene

Ethyl fluoride

Ethylidene fluoride

Ethyl methyl ether, gas or liquid

Ethyne

Hydrocarbon gas, n.o.s.

Hydrogen gas

Isobutane (Liquefied petroleum gas)

Isobutene (Liquefied petroleum gas)

Isobutylene

Laughing gas (Nitrous oxide; Noncombustible)

LPG (Liquefied petroleum gas, Liquefied hydrocarbon gas)

LNG (Liquefied natural gas)

Marsh gas

Methane

Methylacetylene

Methylacetylene-propadiene, stabilized (MAPP)

Methyl chloride (Gas or Liquid)

Methyl ether

Methyl ethyl ether

Methyl fluoride

Methyl vinyl ether (MVE)

Natural gas

Nitrous oxide (Noncombustible)

Oil gas

Perchloryl fluoride

Perfluoroethyl vinyl ether

Perfluoromethyl vinyl ether

Petroleum gas

Propadiene

Propane

Propene (Propylene)

Propyne

Silane (Silicon tetrahydride; Ignites spontaneously in air)

Synthetic natural gas

Tetrafluoroethylene (Perfluoroethylene, TFE)

Trifluoroethane

Tritium-3, gas

Vinyl bromide

Vinyl chloride (VC)

Vinyl fluoride

Vinylidine fluoride

Vinyl methyl ether

RGN 36 FLAMMABLE AND TOXIC GASES

Many of these gases are extremely hazardous, toxic, and flammable. Human exposure to the compounds can be fatal if inhaled or absorbed through the skin. Individuals tend to lose their sense of smell when exposed to many of these compounds. Fire will produce irritating, corrosive, and/or toxic gases. Some gases may be ignited by heat, sparks, or flames and form explosive mixtures in air. A few gases, such as isocyanic acid and methyl nitrite, are a severe explosion risk. Some of the gases will react violently with water and/or oxidizing materials. Gases, such as ethylene oxide, tend to polymerize. Vapors can travel to the source of ignition and flash back. Containers may explode when heated. Most of these gases do not react with liquids. However, in the presence of a spark, these gases may react with liquids in the vapor phase (within specific ranges of concentration) to produce fire, toxic materials, and possibly explosions.

Antimony hydride

Arsine

Blast furnace gas (26% Carbon monoxide, etc.)

Carbon monoxide

Carbon suboxide, gas or liquid

Carbonyl sulfide (Carbonyl sulphide)

Chloroamines (Monochloroamine, Dichloroamine, etc.)

2-Chloropropene

Coal gas

Cobalt hydrocarbonyl, gas

Cyanic acid

Cyanogen, (Gas or Liquid)

Dichloroamine

Dichlorosilane

Ethylene oxide

Fluoromethane

Germane (Germanium tetrahydride)

Hydrogen cyanide (Hydrocyanic acid)

Hydrogen phosphide

Hydrogen selenide

Hydrogen sulfide

Hydrogen telluride

Isocyanic acid

Ketene

Methanethiol

Methyl chlorosilane

Methyl fluoride

Methyl mercaptan

Methyl nitrite

Phosphine

Producer gas

Stibine

Synthesis gas

Trifluorochloroethylene

Water gas (Blue gas)

RGN 37 FLAMMABLE, CORROSIVE, TOXIC AND WATER-SENSITIVE GASES

These gases tend to be extremely flammable and a dangerous fire risk. The compounds can be ignited by heat, sparks, or flames. The gases may form explosive mixtures in air and travel to the source of ignition and flash back. Some compounds, especially diazomethane, are shock sensitive and may react vigorously or explode on contact with alkali metals, rough surfaces, or heat. Containers may explode when heated. Most compounds tend to be toxic and corrosive before and after decomposition. Diazomethane is a known carcinogen (OSHA).

Diazomethane

Diborane (Diboron hexahydride)

Dimethylamine, anhydrous (DMA, anhydrous)

Ethylamine, gas or liquid

Hexahydride diborane

Methylamine, anhydrous

Trimethylamine, anhydrous

RGN 38 OXIDIZING GASES

Oxidizing gases will support combustion and create a fire hazard and/or an explosion. Except for oxygen, most chemicals are fatal if inhaled or absorbed through skin. Some gases react explosively with many types of fuels and organics and may ignite combustibles such as wood, paper, oil, clothing, etc. A number of gases will react and/or explode with air, moist air, contaminates, and/or water. Containers may explode when heated. Many of the gases will react with ammonia, hydrogen, metals, and reducing materials to form toxic gases (deadly) and other toxic chemicals. Fire will produce irritating, corrosive and/or toxic gas. Many of these gases will react with liquids. In the presence of a spark, these gases will react with many liquids in the vapor phase to produce fire, toxic materials, and probably explosions.

Bromine chloride

Chlorine

Chlorine-35,36,37 (Radioactive gas)

Chlorine dioxide

Chlorine monfluoride

Chlorine monoxide

Chlorine pentafluoride

Chlorine trifluoride

Dinitrogen pentoxide

Dinitrogen tetroxide (Nitrogen peroxide)

Fluorine

Fluorine monoxide

Fluorine nitrate, gas or liquid

Nitric oxide

Nitrogen dioxide (Liquid dioxide)

Nitrogen peroxide

Nitrogen tetroxide

Nitrogen trifluoride

Nitrogen trioxide (Nitrogen sesquioxide)

Nitrosyl chloride

Nitrosyl fluoride

Oxygen

Oxygen difluoride

Ozone

Sulfur trioxide (Sulphur trioxide, Sulfuric anhydride)

Tetrafluorohydrazine

RGN 39 TOXIC AND/OR CORROSIVE GASES, INCLUDING COMPRESSED OR LIQUEFIED GASES

The gases may be fatal if inhaled or absorbed through the skin. None ignite readily, but fire will produce irritating, corrosive, and/or toxic gases. Containers may explode when heated. Some of these materials may react violently with water. A few compounds are tear gases or poisonous gases. Some compounds, such as ammonia, will form explosive compounds with specific metals. Ruptured cylinders may rocket. The majority of these gases do *not* react with liquids. However, in the presence of a spark, a few of these gases may react with liquids in the vapor phase (within specific ranges of concentration) during a fire to produce toxic gases.

Ammonia (Ammonium hydroxide concentrated)

Anhydrous ammonia

Artic*

Arsenic pentafluoride

Boron tribromide (Fuming liquid)

Boron trichloride (Fuming liquid)

Boron trifluoride

Bromochlorodifluoromethane

Bromotrifluoromethane

Carbon tetrafluoride

Carbonyl chloride

Carbonyl fluoride

Chlorodifluorobromomethane

Chlorodifluoromethane

Chloropentafluoroacetone

Chloropentafluoroethane

Chlorotetrafluoroethane

Chlorotrifluoroethane

Chlorotrifluoromethane

Cyanogen chloride

Cyanogen fluoride

Dibromoformoxime

Dibromotetrafluoroethane

Dichlorodifluoromethane (Fluorocarbon-12)

Dichlorofluoromethane (Fluorocarbon-21, Dichloromonofluoro-
    methane)

1,2-Dichloro-1,1,2,2-tetrafluoroethane (Fluorocarbon-114)

Dichlorotetrafluoroethane

Diphosgene (Green cross gas)

Fluorodichloromethane

Fluoroform

Fluorotrichloromethane

Formyl fluoride

Freon (Various forms)

Germanium tetrahydride

Halon*, gas

Heptafluoropropane

Hexaethyl tetraphosphate, compressed gas mixture

Hexafluoroacetone

Hexafluoroethane

Hexafluoropropylene

Hexafluoropropylene oxide

Hydrogen bromide, gas, anhydrous

Hydrogen chloride, gas, anhydrous

Hydrogen fluoride, gas, anhydrous

Hydrogen iodide, gas, anhydrous

Methyl bromide

Octafluorobut-2-ene

Octafluorocyclobutane

Octafluoropropane

Pentafluoroethane

Perfluoroisobutylene

Phosgene

Phosphorus pentafluoride

Refrigerants (Refrigerant 112a, Freon 112, etc.)

Selenium hexafluoride

Silicon tetrafluoride

Sulfur dioxide (Sulphur dioxide)

Sulfur hexafluoride (Sulphur hexafluoride)

Sulfur pentafluoride (Disulfur decafluoride)

Sulfur tetrafluoride (Sulphur tetrafluoride)

Sulfuryl fluoride (Sulphuryl fluoride)

Superpalite

Tetrachlorodifluoroethane

Tetrafluoroethane

Tetrafluoromethane

Trichlorofluoromethane (Fluorocarbon-11)

Trifluoroacetyl chloride

Trifluorobromoethane (Halon 1301)

Trifluorobromomethane

Trifluoromethane

Trifluoronitrosomethane

Tungsten hexafluoride

RGN 40 HUMAN CARCINOGENIC COMPOUNDS AND DANGEROUS TOXINS (CHEMICAL AND BIOLOGICAL)

These compounds have been classified as *confirmed* human carcinogens or extremely dangerous to humans. In general, these compounds appear to be carcinogenic and/or fatal when inhaled, ingested, or absorbed through the skin. The amount of time required to directly impact a person's health depends on the compound toxicity level, concentration of the toxicant, level of exposure, and several other variables. These compounds can be identified as human carcinogens and/or OSHA carcinogens in the 14th Edition of Hawley's *Condensed Chemical Dictionary* by R. J. Lewis, Sr. published by John Wiley & Sons, Inc. Some compounds can be identified as human carcinogens by the 1999 TLVs® and BEIs®, "Threshold Limit Values for Chemical Substances and Physical Agents, Biological Exposure Indices," American Conference of Governmental Industrial Hygienists. Additional compounds identified as human carcinogens according to OSHA are usually included in the list. Also, this list does include most chemicals or substances which are *reasonably anticipated to be human carcinogens*. "Reasonably Anticipated to Be Human Carcinogens" included in both the table (Alphabetical List of Compounds and RGNs) and this list (RGN 40) are noted by (*RAHC*) after the chemical name. There are a number of agents, chemicals, and substances appearing on several of the lists identified above. In the future, other chemicals and substances will be identified as human carcinogens that are not included in these lists. For additional information on exposure limits, refer to the appropriate references provided above and the OSHA literature, the U.S. EPA Web site on the Internet, and the International Agency for Research on Cancer (IARC). The report on carcinogens is mandated by Section 301 (b) (40) of the Public Health Services Act and is publishd annually. In addition, a number of communicable diseases common to man are identified as "extremely dangerous biological substances" in the list. This information can be found in the 15th Edition of "Control of Communicable Diseases in Man," Abram S. Benenson, Editor, published by the American Public Health Association, 1990. For specific information about diseases and biological toxins, refer to this reference. For updates and changes, the literature must be periodically reviewed.

*NOTE: A number of the **biological toxins** and "potentially deadly" materials can be rendered ineffective by chlorox solution, 5-6% hydrogen peroxide solution, chlorine dioxide in air, chlorine vapors in air, or by most strong oxidizing gases and solutions. Some **biological toxins** can be dispersed as fine particulates (suspended) in air or as a dry powder deposited on surfaces, while others are transmitted by a bite (mosquito, fleas, etc.) or personal contact with infected animals or humans. Small concentrations of various toxins may exist as deposits on the surface of wool, hair, bone, skin, etc. Many of these toxins require special isolation, handling, and identification techniques using safety equipment and appropriate procedures. Biological toxins are listed in **italics** below. In the case of **chemical toxins**, removal and proper disposal of the toxin and/or chemical dilution of the toxin with an inert substance can prevent or minimize problems. Some chemical toxins can be "detoxified" by reaction with oxidizing agents, or reducing agents, or other "deactivating" chemicals. Not every biological toxin or cancer-causing material is included in this list*

2-Acetylaminofluorene (*RAHC*)

Acrylamide (*RAHC*)

Acrylonitrile (*RAHC*)

Aflatoxins, B_2 and G_1 strains

Agent orange (Traces of chlorinated dioxins)

2-Aminoanthraquinone (*RAHC*)

o-Aminoazotoluene (*RAHC*)

4-Aminodiphenyl (4-Aminobiphenyl, *p*-Aminodiphenyl)

1-Amino-2-methylanthraquinone (*RAHC*)

2-Aminonaphthalene (2-Naphthylamine)

Amitrole (*RAHC*)

Ammonium cadmium chloride

Ammonium chromate

Ammonium dichromate

Ammonium peroxychromate

o-Anisidine hydrochloride (*RAHC*)

Anthrax (*Bacillus anthracis*; Acute bacterial disease; Dry powder; Very fine powder suspended in air; Found on animal hides, wool, hair, bones, etc.)

Antimony trioxide

Aroclor® 1254 (*RAHC*)

Aroclor® 1260 (*RAHC*)

Arsenic

Arsenic compounds, inorganic (Arsenic trioxide, etc.)

Asbestos dust particles

Azacitidine (*RAHC*)

Barium chromate

BCNU (*RAHC*)

Benz(*e*)acephenanthrylene

Benz[*a*]anthracene (*RAHC*)

Benzo[*b*]fluoranthene (*RAHC*)

Benzo[*j*]fluoranthene (*RAHC*)

Benzo[*k*]fluoranthene (*RAHC*)

Benzo(*b*)phenanthrene

Benzene

Benzene hexachloride (*RAHC*)

Benzidine

Benzo[*b*]fluoranthene

Benzo[*a*]pyrene (Benzopyrene) (*RAHC*)

Benzo[*e*]pyrene

Benzotrichloride (*RAHC*)

Beryl or Beryl ore (*RAHC*)

Beryllium

Beryllium aluminum alloy (*RAHC*)

Beryllium chloride (*RAHC*)

Beryllium compounds (Beryllium fluoride, Beryllium sulfate, etc.)

Beryllium dust

Beryllium fluoride (*RAHC*)

Beryllium hydroxide (*RAHC*)

Beryllium oxide (*RAHC*)

Beryllium phosphate (*RAHC*)

Beryllium sulfate (*RAHC*)

Beryllium sulfate tetrahydrate (*RAHC*)

Beryllium zinc silicate (*RAHC*)

BHC (*RAHC*)

Bis(chloromethyl)ether

Bis(chloroethyl) nitrosourea (BCNU; *RAHC*)

Bis(dimethylamino)benzophenone (*RAHC*)

Bis(2-ethylhexyl) phthalate (*RAHC*)

Bromodichloromethane (*RAHC*)

Bubonic plague (Bacterial infection; Infected fleas and wild rodents, infected pets, person-to-person)

Busulfan

1,3-Butadiene

1,4-Butanediol dimethylsulfonate

Cadmium, fume, dust, or very fine powder

Cadmium compounds (Cadmium chloride, Cadmium oxide, Cadmium sulfate, Cadmium sulfide, etc.)

Calcium chromate

Camphechlor (*RAHC*)

Captab (*RAHC*)

Captan (*RAHC*)

Carbon tetrachloride (*RAHC*)

CCNU (*RAHC*)

Chlorambucil

Chlorobenzilate* (Highly suspected carcinogen)

Chlordecon (*RAHC*)

Chlorendic acid (*RAHC*)

Chloroethylene

1-(2-Chloroethyl)-3-(4-methylcyclohexyl)-1-nitrosourea (MeCCNU)

1-(2-Chloroethyl)-3-cyclohexyl-1-nitrosourea (*RAHC*)

Chlorinated paraffins (C_{12}, 60% Chlorine; *RAHC*)

Chloroform (*RAHC*)

Chloromethyl ether as Bis(Chloromethyl) ether

Chloromethyl methyl ether

3-Chloro-2-methylpropene (*RAHC*)

4-Chloro-*o*-phenylenediamine (*RAHC*)

Chloroprene (*RAHC*)

p-Chloro-*o*-toluidine (*RAHC*)

p-Chloro-*o*-toluidine hydrochloride (*RAHC*)

Chromic acid

Chromium VI compounds, soluble and insoluble (Chromic acid, Chromium trioxide, etc.)

Coal tar pitch volatiles (Benzene solubles)

Coke oven emissions

Creosote (Coal, Wood, Tars, etc.)

p-Cresidine (*RAHC*)

Cristobalite

Cupferron (*RAHC*)

Cyclophosphamide

Cyclosporin A (Cyclosporine A, Ciclosporin)

Danthron (*RAHC*)

Daunomycin (Daunorubicin)

DDT (*RAHC*)

Decabromobiphenyl (*RAHC*)

DEHP (*RAHC*)

DEN (*RAHC*)

Diamine

2,4-Diaminoanisole sulfate (*RAHC*)

Diaminodiphenyl ether (*RAHC*)

4,4'-Diaminodiphenylmethane (*p,p*'-Diaminodiphenylmethane, MDA)

2,4-Diaminotoluene (*RAHC*)

Diazomethane

Dibenz[*a,h*]acridine (*RAHC*)

Dibenz[*a,j*]acridine (*RAHC*)

Dibenz[*a,h*]anthracene (*RAHC*)

7*H*-Dibenzo[*c,g*]carbazole (*RAHC*)

Dibenzo[*a,e*]pyrene, Dibenzo[*a,h*]pyrene, Dibenzo[*a,i*]pyrene and Dibenzo[*a,l*]pyrene (Several listings of each; *RAHC*)

1,2-Dibromo-3-chloropropane (*RAHC*)

1,2-Dibromoethane (*RAHC*)

1,4-Dichlorobenzene (*p*-Dichlorobenzene; *RAHC*)

3,3'-Dichlorobenzidine (*RAHC*)

1,4-Dichloro-2-butene

Dichlorodiethyl sulfide (Mustard gas)

Dichlorodiphenyltrichloroethane (*RAHC*)

1,2-Dichloroethane (*RAHC*)

Dichloromethane (*RAHC*)

1,3-Dichloropropene (Technical grade; *RAHC*)

Diepoxybutane (*RAHC*)

Diesel exhaust particulates (*RAHC*)

N,N-Diethyldithiocarbamic acid 2-chloroallyl ester (*RAHC*)

Di(2-ethylhexyl) phthalate (*RAHC*)

Diethylnitrosamine (*RAHC*)

Diethylstilbestrol (DES)

Diethyl sulfate (*RAHC*)

Diglycidyl resorcinol ether (*RAHC*)

1,8-Dihydroxyanthraquinone (*RAHC*)

3,3'-Dimethoxybenzidine (*RAHC*)

4-Dimethylaminoazobenzene (*RAHC*)

3,3'-Dimethylbenzidine (*RAHC*)

Dimethylcarbamoyl chloride (*RAHC*)

1,1-Dimethylhydrazine (UDMH; *RAHC*)

Dimethylnitrosamine (*RAHC*)

Dimethyl sulfate (Protective clothing required; *RAHC*)

Dimethylvinyl chloride (*RAHC*)

1,6-Dinitropyrene (*RAHC*)

1,8-Dinitropyrene (*RAHC*)

1,4-Dioxane (*RAHC*)

DMN (*RAHC*)

Ebola (Virus, Transmitted by contact)

ENU (*RAHC*)

Environmental tobacco smoke

Epichlorohydrin (*RAHC*)

Erionite

Ethyl carbamate (*RAHC*)

Ethylene dibromide (*RAHC*)

Ethylene dichloride (*RAHC*)

Ethylene oxide

Ethylene thiourea (*RAHC*)

Ethyl methanesulfonate (*RAHC*)

N-Ethyl-*N*-nitrosourea (*RAHC*)

FireMaster BP-6 (*RAHC*)

FireMaster FF-1 (*RAHC*)

Formaldehyde (Vapors; Nasal)

HCCH (HCH, *RAHC*)

Hematite, red

Hexabromobiphenyl (*RAHC*)

Hexachlorobenzene (*RAHC*)

Hexachlorocyclohexane (α-, β-, γ- forms, *RAHC*)

Hexachloroethane (*RAHC*)

Hexamethylphosphoramide (Hempa; *RAHC*)

Hydrazine (*RAHC*)

Hydrazine sulfate (*RAHC*)

Hydrazobenzene (*RAHC*)

Indeno[1,2,3—*cd*]pyrene (*RAHC*)

Isotox* (*RAHC*)

Kanechlor® (*RAHC*)

Kepone® (*RAHC*)

Lead acetate (*RAHC*)

Lead chromate

Lead naphthenate

Lead phosphate (*RAHC*)

Lindane (*RAHC*)

MBOCA (MOCA®) (*RAHC*)

Mechlorethamine hydrochloride (Vesicant, Burns/Blisters)

Melphalan

2-Methylaziridine (*RAHC*)

5-Methylchrysene (*RAHC*)

4,4'-Methylenebis(2-chloroaniline) (*RAHC*)

4,4'-Methylenebis(N,N-dimethylbenzenamine) (*RAHC*)

Methylene chloride (*RAHC*)

4,4'-Methylenedianiline (*RAHC*)

4,4'-Methylenedianiline dihydrochloride (*RAHC*)

Methyl methanesulfonate (*RAHC*)

N-Methyl-*N*'-nitro-*N*-nitrosoguanidine (*RAHC*)

N-Methyl-*N*-nitrosourea (*RAHC*)

Methyl-*p*-toluenesulfonate (Vesicant, Blisters, and burns)

Michler's ketone (*RAHC*)

Mirex (*RAHC*)

Mustard gas

Myleran®

β-Naphthylamine (2-Naphthylamine)

Nickel (Dust; *RAHC*)

Nickel acetate (*RAHC*)

Nickel carbonate (*RAHC*)

Nickel carbonyl (Nickel tetracarbonyl; *RAHC*)

Nickel compounds, insoluble (Dust; Lung cancer)

Nickel hydroxide (*RAHC*)

Nickelocene (*RAHC*)

Nickel oxide (*RAHC*)

Nickel subsulfide (Lung cancer; *RAHC*)

Nitrilotriacetic acid (*RAHC*)

o-Nitroanisole (*RAHC*)

6-Nitrochrysene (*RAHC*)

4-Nitrodiphenyl (4-Nitrobiphenyl, o-Nitrobiphenyl, ONB)

Nitrofen (*RAHC*)

Nitrogen mustard hydrochloride (Several compounds; *RAHC*)

2-Nitropropane (*RAHC*)

1-Nitropyrene (*RAHC*)

4-Nitropyrene (*RAHC*)

N-Nitrosodi-*n*-butylamine (*RAHC*)

N-Nitrosodiethanolamine (*RAHC*)

N-Nitrosodiethylamine (*RAHC*)

N -Nitrosodimethylamine (*RAHC*)

N-Nitrosodi-n-propylamine (*RAHC*)

4-(N-Nitrosomethylamino)-1-(3-pyridyl)-1-butanone (*RAHC*)

N-Nitrosomethylvinylamine (*RAHC*)

N-Nitrosomorpholine (*RAHC*)

N-Nitrosonornicotine (*RAHC*)

N-Nitrosopiperidine (*RAHC*)

N-Nitrosopyrrolidine (*RAHC*)

N-Nitrososarcosine (*RAHC*)

Octabromobiphenyl (*RAHC*)

4,4'-Oxydianiline (*RAHC*)

PAHs (Polycyclic aromatic hydrocarbons; *RAHC*)

PBBs (Polybrominated biphenyls; *RAHC*)

PCBs (Polychlorinated biphenyls; *RAHC*)

Perchloroethylene (Tetrachloroethylene; *RAHC*)

Phenacetin (*RAHC*)

Phenazopyridine hydrochloride (*RAHC*)

Phenoxybenzamine hydrochloride (*RAHC*)

Pneumonic plague (Bacterial infection of lung; person-to-person or infected animal; Transmitted by aerosolized droplets)

Polybrominated biphenyls (PBBs; *RAHC*)

Polychlorcamphene (*RAHC*)

Polychlorinated biphenyls (PCBs; *RAHC*)

Polycyclic aromatic hydrocarbons (PAHs; *RAHC*)

Procarbazine hydrochloride (*RAHC*)

1,3-Propane sultone (*RAHC*)

β-Propiolactone (*RAHC*)

Propylene oxide (*RAHC*)

Propylenimine (*RAHC*)

Propylthiouracil (*RAHC*)

Radon (Radon-222)

Reserpine (*RAHC*)

Safrole (*RAHC*)

Selenium sulfide (*RAHC*)

Silica (Tridymite; Very fine crystalline or respirable size)

Smallpox (Variola virus ; Two known types; Personal contact or close contact through air)

Strong Inorganic Acid Mists Containing Sulfuric Acid

Strontium chromate

Sulfallate (*RAHC*)

Sulfuric acid, concentrated, mists (Throat, Larynx)

TBH (*RAHC*)

2,3,7,8-Tetrachlorodibenzo-*p*-dioxin (TCDD, *RAHC*)

Tetrachloroethylene (*RAHC*)

Tetranitromethane (*RAHC*)

Thioacetamide (*RAHC*)

Thiotepa

Thiourea (*RAHC*)

Thorium dioxide

Tobacco smoking (Chronic inhalation of smoke)

Tok* (Nitrofen, *RAHC*)

o-Toluidine (*RAHC*)

o-Toluidine hydrochloride (*RAHC*)

Toxaphene (*RAHC*)

Toxaphene with methyl parathion (*RAHC*)

Trichloroethylene (*RAHC*)

2,4,6-Trichlorophenol (*RAHC*)

1,2,3-Trichloropropane (*RAHC*)

Triethylenethiophosphoramide

Tris(1-aziridinyl)phosphine sulfide

Tris(2,3-dibromopropyl) phosphate (*RAHC*)

UDMH (*RAHC*)

Uranium (Natural, Pyrophoric metal)

Uranium-233,234,235,238 (Isotopes; Metal, alloys, Soluble)

Uranium-233,234,235,238 compounds, n.o.s. (Soluble and insoluble)

Urethane (Urethan, Ethyl carbamate; *RAHC*)

VEE or *Venezuelan Equine Encephalitis* (Infected mosquito; Aerosol infection; probably not person to person)

VC (Vinyl chloride)

Vinyl bromide

Vinyl chloride

Vinyl cyanide (*RAHC*)

4-Vinyl-1-cyclohexene diepoxide (*RAHC*)

Vinyl fluoride

Zeidane (*RAHC*)

Zinc chromate

RGN 61 COMBUSTIBLE AND FLAMMABLE MATERIALS, MISCELLANEOUS

These compounds can be highly flammable, flammable, or combustible. When heated, the compounds can *evolve fumes* that are irritants, poisonous, and/or toxic. Many of the compounds are explosive in acid. Some compounds can be ignited by friction. A number of substances are a fire and explosion risk. At times, some chemicals may re-ignite after the fire is extinguished. A number of substances will pressurize the container during heating. Avoid breathing toxic gases formed during high temperature reaction, volatilization, or decomposition.

Alkyl resins, n.o.s.

Ammonium acetate

Ammonium benzoate

Ammonium bicarbonate

Ammonium bisulfite (Ammonium bisulphite)

Ammonium bromide

Ammonium carbamate

Ammonium carbonate

Ammonium chloride

Ammonium citrate

Ammonium hydrogen sulfate

Ammonium oxalate

Ammonium sulfamate

Ammonium tartrate

Amyl acid phosphate

Anthraquinone

Asphalt (petroleum) fumes

Bakelite*

S-Bioallethrin

Borneol (Bornyl alcohol)

Calcium dodecylbenzenesulphonate

Calixin*

2-Camphanol

Camphene

Camphor

Camphor oil

Cellulose

Cholecalciferol

Cocculus (Cocculin)

Diesel exhaust particlates

Diesel fuel (Kerosene)

Dimehypo

Esbiol*

Esdepalléthrine

Fuel oil (no.1 through no. 6)

Fuels (Gasoline, Aircraft fuels, Jet fuel, Diesel fuel, etc.)

Gasoline (Gasohol, petrol, etc.)

Gas oil

Gemini*

8-Hydroxyquinoline (Oxine)

Kerosene

Lacquer

Minerol oil, vapor, or particulates

Mineral spirits

Naphtha (Spirits)

Natural gasoline

Oil, petroleum, n.o.s. (Mist)

Petrol

Petroleum crude oil

Petroleum distillates (Paints, Lacquer, Paint oils, Paint thinners, Grease, and similar substances)

Petroleum hydrocarbons (Kerosene, Fuel oil, Gasoline, Turpentine, Engine or motor oils, Gear box oils, Jet oil, Transmission oils, Petroleum oil, etc.)

Petroleum naptha

Petroleum spirits (Naphtha, Spirits, Petroleum ether)

Picrotoxin

Polyacrylamide polymer decomposition products

Polyacrylonitrile decomposition products (Orlon, Dynel, etc.; High temperature stability)

Polyamide polymer decomposition products

Polyurethane decomposition products

Quinoline

Remlat*

Shale oil

Sheliflex*

Stoddard solvent

Sulfolane

Synfuel

Synthetic rubber decomposition products (Not containing sulfur)

Tridemorph

Turpentine

Turpentine substitutes

Unisolve

Varsol*

VM & P Naphtha

RGN 62 EXPLOSIVE HAZARD AND SHOCK-SENSITIVE COMPOUNDS

Most of these compounds are shock sensitive and/or explosive. Many of the compounds can be defined as "high explosives" or a "severe explosion hazard." Many of the compounds are heat sensitive, acid and/or oxygen sensitive, shock sensitive, and strong oxidizing agents. Some toxic compounds are extremely flammable forming a variety of toxic gases. A number of compounds ignite and/or explode spontaneously when dry or when heated slightly. Additional compounds may explode at room temperature or under vacuum. Some compounds are identified as dangerous fire and/or explosion hazards. A number of oxidizers can be listed in this group and/or in group RGN 64. Some oxidizers may ignite oil, wood, paper, cloth, and many organic materials. Many compounds with relatively low explosion limits are not identified in this list. Most oxidizers are not included in this list. Avoid breathing toxic vapors of the chemicals and the decomposition and/or reaction products. Experienced individuals should handle these chemicals properly.

Each compound is identified by an explosion-type category. Often a fire occurs before an explosion occurs. However, a fire can occur *before*, *after,* or *during* an explosion depending on air content, chemical composition, flame or spark presence, environmental conditions, and many other factors. Most explosive compounds require a specific procedure or physical condition for detonation (explosion). Some "potentially" explosive compounds are listed in RGN 28, RGN 27, RGN 14, or in other groups. **Note**: *Not every explosive compound or shock sensitive compound is identified in this list.*

Definitions:

E = severe explosion hazard (heat, shock, dry, friction, rough surfaces, etc.)

EV = explode under vacuum

S = dangerous explosion risk or may explode (shock and/or impact sensitive, heat and/or flame sensitive, temperature sensitive, light sensitive, pressure sensitive, contamination sensitive, and/or vibration and friction sensitive); fire risk or severe fire risk with other compounds (e.g., severe fire risk with compounds such as peroxides in ethers, silver salts in organics, etc.)

X = dangerous (or severe) fire and explosion hazard (moderate to severe) if shocked, heated, friction, or if sparks are present.

Acetyl azide (S)

Acetyl nitrate (E)

Acetyl peroxide, pure material (E)

Aerozine (UDMH; X)

Aluminum picrate (E)

Amatol (E)

2-Amino-4,6-dinitrophenol (E)

Ammonium azide (S)

Ammonium chlorate (S)

Ammonium hexanitrocobaltate (E)

Ammonium nitrate (X)

Ammonium perchlorate (S)

Ammonium periodate (S)

Ammonium permanganate (S)

Ammonium picrate (E)

Ammonium tetraperoxychromate (S)

Anisoyl chloride (S)

Aspirin dust (S)

Azide, n.o.s. (S)

Azidocarbonyl guanidine (E)

Barium azide (E)

Benzenediazonium chloride (E)

Benzotriazole (EV)

Black powder (X)

Boron triazide (E)

Bromine azide (S)

Bromopicrin (E)

Butadiene (S, May form explosive peroxides in air)

Butadiyne (S)

Butanetriol trinitrate (E)

Butyl ether, anhydrous with peroxides (X)

t-Butyl hypochlorite (S)

t-Butylperoxypivalate (X)

Cadmium azide (E)

Cadmium nitrate (X)

Cadmium nitride (S)

Calcium nitrate (S)

Carbon disulfide (Carbon bisulfide) (X)

Cellulose nitrate (X)

Cesium azide (S)

Chlorine azide (S)

Chlorine dioxide, hydrate (Frozen) (S)

Chlorine heptoxide (S)

Chlorine trioxide (S)

Chloroacetylene (E)

Chloronitropropane (X)

Chlorotrinitrobenzene (2-Chloro-1,3,5-trinitrobenzene) (E)

Chromic nitrate (Chromium nitrate) (S)

Colloidion (With impurities) (X)

Copper acetylide (Cuprous acetylide) (S)

Cuprous acetylide (E)

Cyanic acid (E)

Cyanogen azide (E)

Cyanuric triazide (E)

Cyclonite (E)

Cyclotrimethylenetrinitramine (E)

DDNP (S)

Decaborane (X)

DEGN (E)

Diacetylene (S)

Diamine (E)

para-Diazobenzenesulfonic acid (S)

Diazodinitrophenol (S)

Diazomethane (E)

1-Diazo-2-naphthol-4-sulfonic acid (S)

Dibromoacetylene(s) (E)

Dicopper(I) acetylide (S)

Dicyclohexylammonium nitrite (X)

Dicyclohexyl peroxydicarbonate (S)

Diethylaluminum chloride (X)

Diethylaluminum hydride (X)

Diethylene glycol dinitrate (E)

Dimethoxyethane (X)

Dimethyl cadmium, pyrophoric (S)

Dimethyl ether, compressed gas or liquid (S)

Dimethyl sulfide (X)

Dinitroaminophenol (S)

Dinitrobenzene (S)

Dinitronaphthalene (X)

Dinitrophenol, dry (E)

Dinitrophenylhydrazine (S)

Dinitroresorcinol (S)

2,4-Dinitrosoresorcinol (E)

1,4-Dioxane with peroxides (S)

Dipentaerythritol hexanitrate (E)

Dipicrylamine (E)

Dipicryl sulfide (E)

Erbium nitrate (S)

Erythrityl tetranitrate (Erythrol tetranitrate) (E)

Ethers (Low molecular weight ethers with explosive peroxides) (E)

Ethyl acetate (X)

Ethyl acrylate (X)

Ethyleneimine (X)

Ethyl ether containing peroxides (E)

Ethyl methacrylate (X)

Ethyl methyl ether (S)

Ethyl nitrate (X)

Ethyl nitrite (S, X)

Explosives, n.o.s. (E, S)

Fluorine azide (E)

Fluorine nitrate, liquid (S)

Glycol dinitrate (E)

Glycol monolactate trinitrate (E)

Gold fulminate (S)

Guanidine nitrate (S)

Guanyl nitrosaminoguanylidene hydrazine (E)

Gun cotton (X)

Hexanitrodiphenylamine (Hexite, Hexil) (E)

Hexanitrodiphenyl sulfide (E)

HMX (E)

Hydrazine (Anhydrous) (E)

Hydrazine hydrate (E)

Hydrazine nitrate (E)

Hydrazine perchlorate (E)

Hydrazoic acid (S)

Hydrogen azide (S)

Isocyanic acid (Cyanic acid) (E)

Lead azide (E)

Lead dinitroresorcinate (E)

Lead mononitroresorcinate (S)

Lead nitride (E)

Lead styphnate (E)

Lead trinitroresorcinate (E)

Mannitol hexanitrate (E)

Mercuric cyanate, dry (E)

Mercuric oxycyanide (S)

Mercurous acetylide (E)

Mercury fulminate, dry (E)

Mercury nitride (E)

Methyl acetate (X)

Methyl acrylate (X)

Methyl nitrate (E)

Methyl nitrite (E)

Methyl picrate (E)

Naphite (S)

Naphthacene (S)

Nitramine (X)

Nitrobromoform (E)

Nitro carbo nitrate (E)

Nitrocellulose, dry (X)

Nitrogen trichloride (Explodes in direct sunlight; S)

Nitrogen trifluoride (E)

Nitrogen triiodide, dry (E)

Nitroglycerin (E)

Nitroguanidine (E)

Nitromannite (E)

Nitromethane (X)

ortho-Nitrophenylpropiolic acid (S; During decomposition only)

Nitrosoguanidine (E)

para-Nitrosophenol (E, X)

Nitrostarch, dry (E) or almost dry (X)

Nitrourea (E)

Pentaerythritol tetranitrate (E)

Pentolite (E)

Performic acid

PETN (E)

Phenol trinitrate (E)

Picramic acid (Picraminic acid) (S, X)

Picramide (E)

Picric acid (E)

Picrite, dry (S)

Picryl chloride (E)

Polyvinyl nitrate (X)

Potassium azide (S)

Potassium styphnate (E)

Propyl nitrate (X)

Pyroxylin (X)

RDX (E)

Silver acetylide (E)

Silver azide (E)

Silver nitride (E)

Silver picrate (E)

Silver styphnate (E)

Silver tetrazene (E)

Silver trinitromethanide (E)

Silver trinitroresorcinate (E)

Smokeless powder (X)

Sodium azide (S)

Sodium picramate, dry (X)

Starch nitrate, dry (E) or almost dry (X)

Styphnic acid (E)

TATP (X)

Tetranitroaniline (X)

Tetranitromethane (X)

Tetrasilane (X)

Tetrasulfur tetranitride (E)

Tetrazene (E)

Tetryl (X)

Thallium nitride (S)

TNA (X)

TNB (E)

TNT (S, X)

Trilead dinitride (E)

Trimercury dinitride (E)

Trinitroaniline (S)

Trinitroanisole (E, S)

1,3,5-Trinitrobenzene (S)

Trinitrobenzoic acid (2,4,6- Trinitrobenzoic acid) (E)

2,4,6-Trinitro-*meta*-cresol (E)

Trinitroglycerin (E)

Trinitromethane (S)

Trinitronaphthalene (S)

Trinitrophenol (E)

2,4,6-Trinitrophenyl methyl ether (E, S)

Trinitrophenylmethylnitramine (X)

2,4,6-Trinitroresorcinol (E)

2,4,6-Trinitrotoluene (S, X)

Trinitrotrimethylenetriamine (E)

Urea nitrate (X)

Vinyl azide (S)

Xenon trioxide, dry (S)

Zirconium picramate, dry (S)

RGN 63 POLYMERIZABLE COMPOUNDS

These compounds are either flammable or combustible. Some compounds can polymerize explosively, while others polymerize when heated or when exposed to light or moisture. Substances may be ignited by heat, sparks, and flames. In a fire, these compounds may evolve flammable hydrogen. Vapors may travel to the source of ignition and flash back. Vapors can form explosive mixtures with air. Many compounds can be chemically inhibited to avoid or minimize polymerization.

Acrolein

Acrylamide

Acrylic acid

Acrylonitrile (Human carcinogen)

Allyltrichlorosilane

Aqualin

Aziridine

Bicyclo[2.2.1]hepta-2,5-diene

1,3-Butadiene (Human carcinogen)

Butadiyne

n-Butyl acrylate

Carbon suboxide (Lachrymator)

1-Chloro-2,3-epoxypropane

Chloroethylene (Human carcinogen)

Chloroprene

Chloropropylene oxide

Chromium oxyfluoride

Chromyl fluoride

Crotonaldehyde (Lachrymator)

3-Cyclooctadiene

Cyclooctadiene(s)

Cyclooctatetraene

Diallyl ether

Diallyl maleate

Difluoroethylene

Diketene (Acetyl ketene)

Dimethylaminoethyl methacrylate (Lachrymator)

DVB, uninhibited (Vinylstyrene)

Divinylbenzene, uninhibited (DVB)

Divinyl sulfide

Epichlorohydrin

Epoxyethane

Ethene

Ethylacetylene

Ethyl acrylate

Ethylene

Ethyleneimine (Carcinogen)

Ethylene oxide (Human carcinogen)

2-Ethylhexyl acrylate

Ethyl methacrylate

Furaldehyde(s)

Furfural

Furfuraldehyde

Glycidaldehyde

Hydrocyanic acid (Prussic acid; Unpure, Not stabilized)

Hydrogen cyanide liquid (Unpure, Not stabilized)

Indene

Isobutyl acrylate

Isobutyl methacrylate

Isoprene

Isopropenyl acetate

IVE

Ketene

Methacrylaldehyde (Methacrolein)

Methacrylic acid

Methacrylonitrile

Methyl acrylate

Methylallyl alcohol

Methylallyl chloride (MAC)

Methyl butadiene

Methyl isopropenyl ketone

Methyl methacrylate

2-Methyl styrene

Methyl vinyl ether (MVE)

Methyl vinyl ketone

2,5-Norbornadiene

Phenylethylene (Styrene monomer)

Polymerizable compound, n.o.s.

Propadiene

2-Propenal

Propylene oxide

Styrene, monomer

Tetrafluoroethylene (Perfluoroethylene, TFE)

Triethylenemelamine (TEM)

Trifluorochloroethylene

VC (Human carcinogen)

Vinyl acetate

Vinyl azide

Vinylbenzene

Vinyl bromide (Human carcinogen)

Vinyl butyrate

Vinyl chloride (Human carcinogen)

Vinyl cyanide

Vinyl ethyl ether (EVE)

Vinyl fluoride (Human carcinogen)

Vinylidene chloride

Vinylidene fluoride

Vinyl isobutyl ether (IVE)

Vinyl isopropyl ether

Vinyl methyl ether (MVE)

Vinyl pyridine(s)

Vinyltoluene

Vinyl trichlorosilane

RGN 64 OXIDIZING AGENTS, OXIDIZERS (STRONG TO MEDIUM STRENGTH) AND SELECTIVE OXIDIZERS

These compounds may ignite with combustibles such as oil, paper, wood, cloth, fuels, and other hydrocarbons. Some of the inorganic oxidizing agents will explode or react vigorously when in contact with organic (combustible) materials, reducing agents, and/or specific metal compounds. In addition, chlorox-type solutions mixed with aqueous solutions of acids (i.e., hydrochloric acid) will often produce a toxic gas. These chemicals in contact with metals may evolve toxic fumes and flammable gases. Do not mix these chemicals with reducing agents (RGN 65). Many of the compounds are heat and shock sensitive in acid. A few compounds can detonate when exposed to heat, light, vibration, shock, or other effects. Avoid breathing the vapors of these compounds and the products of reaction and/or decomposition. Experienced individuals should handle these chemicals properly. A number of oxidizing agents included in this list have limited application and/or limited strength as oxidizing agents. **Note:** *The oxidizing agents, organic peroxides, and organic hydroperoxides, are presented in RGN 30. Various types of organic oxidizing agents are listed in RGN 64. Most organic oxidizing agents are not as reactive or strong as inorganic oxidizing agents. Not every oxidizing agent is identified in this list.*

Aluminum chlorate

Aluminum nitrate

Aluminum picrate

Ammonium chlorate

Ammonium dichromate

Ammonium dinitro-*o*-cresolate

Ammonium iodate

Ammonium nitrate

Ammonium perchlorate

Ammonium periodate

Ammonium permanganate

Ammonium peroxychromate

Ammonium persulfate

Ammonium sulfate nitrate

Ammonium tetrachromate

Ammonium tetraperoxychromate

Ammonium trichromate

Amyl nitrate

Amyl nitrite

Antimony perchlorate

Aqua regia

Ascaridole

Barium bromate

Barium chlorate

Barium hypochlorite

Barium iodate

Barium nitrate

Barium nitrite

Barium perchlorate

Barium permanganate

Barium peroxide

Beryllium nitrate

Bismuth nitrate

Bleach (Hydrogen peroxide, Sodium hypochlorite, Sodium peroxide, Sodium chlorite, Calcium hypochlorite, Hypochlorous acid, Chlorinated lime, Sodium perborate, 1,3-Dichloro-5,5-dimethylhydantoin [DDH], and organic-chlorine derivatives)

Bleach liquor (Calcium hypochlorite)

Borate and chlorate mixtures

Bromates, n.o.s.

Bromic acid

Bromine

Bromine azide

Bromine chloride

Bromine monofluoride

Bromine pentafluoride

Bromine trifluoride

3-Bromo-1-chloro-5,5-dimethylhydantoin (Specific applications)

t-Butyl chromate

t-Butyl hypochlorite

Butyl nitrate (*tert*-, *iso*- *n*-)

Cadmium bromate

Cadmium chlorate

Cadmium iodate

Cadmium nitrate

Caesium nitrate

Caesium nitrite

Calcium ammonium nitrate

Calcium bromate

Calcium chlorate

Calcium chlorite

Calcium chromate

Calcium dichromate

Calcium hydrogen sulfite (Specific applications)

Calcium hypochlorite
Calcium iodate
Calcium nitrate
Calcium oxychloride
Calcium perchlorate
Calcium perchromate
Calcium permanganate
Calcium peroxide
Calcium plumbate
Caro's acid (Peroxysulfuric acid)
Ceric ammonium nitrate
Ceric sulfate
Cerium nitrate
Cerous nitrate
Cesium nitrate (Caesium nitrate)
Cesium nitrite (Caesium nitrite)
Cesium perchlorate (Caesium perchlorate)
Cesium peroxide
Cesium tetroxide
Cesium trioxide
Chloramine-T
Chlorate, inorganic, n.o.s.
Chloric acid
Chlorine, gas
Chlorine-35,36,37 (Radioactive gas)
Chlorine (In basic solution)
Chlorine dioxide
Chlorine dioxide, hydrate (Frozen)
Chlorine heptoxide
Chlorine monofluoride

Chlorine monoxide

Chlorine pentafluoride

Chlorine trifluoride

Chlorine trioxide

Chlorine water (Up to 0.4%)

Chloriodized oil (Mixed with water)

Chlorites, inorganic, n.o.s.

Chlorochromic anhydride

N-Chlorosuccinimide (NCS)

Chlorox (Sodium/Potassium hypochlorite)

Chromic acid

Chromic anhydride

Chromic nitrate (Chromium nitrate)

Chromium oxychloride

Chromium trioxide

Chromyl chloride

Cobaltous nitrate (Cobalt nitrate)

Cobaltous perchlorate

Copper chlorate

Copper nitrate (Cupric nitrate)

Cyanuric acid

DDH

1,3-Dibromo-5,5-dimethylhydantoin

Di-*tert*-butylquinone (Oxidant, Polmerization catalyst)

Dichloroamine-T

Dichloroamine

1,3-Dichloro-5,5-dimethylhydantoin [DDH]

Dichloroisocyanuric acid

Dichloroisocyanuric acid salts

Dichloro-*s*-triazine-2,4,6-trione

Didymium nitrate

Dinitrogen pentoxide

Dinitrogen tetroxide

Dysprosium nitrate

Ethylene chromic oxide

Ferric chromate

Ferric dichromate

Ferric nitrate

Fluorine

Fluorine monoxide

Fluorine nitrate

Gadolinium nitrate

Guanidine nitrate

Halazone

HTH

Hydrazine nitrate

Hydrazine perchlorate

Hydrogen peroxide

Hydrolin* (Ammonium nitrate solutions)

Hydrozon*

Hypobromous acid

Hypochlorite solution, n.o.s.

Hypochlorites, inorganic, n.o.s.

Hypochlorous acid

Iodic acid anhydride

Iodine pentoxide

Isopropyl nitrate (2-propanol nitrate)

Javel water

Kastone*

Kjelgest*

Lanthanum ammonium nitrate

Lanthanum nitrate

Lead chlorite

Lead dioxide

Lead nitrate

Lead oxide

Lead perchlorate

Lead peroxide

Lead tetraacetate

Lime, chlorinated

Lithium chlorate

Lithium chromate

Lithium hypochlorite

Lithium iodate

Lithium nitrate

Lithium perchlorate

Lithium peroxide

Liquid dioxide

Lo-Bax*

Lutetium nitrate

Magnesium bromate

Magnesium chlorate

Magnesium chloride and chlorate mixture

Magnesium nitrate

Magnesium perborate

Magnesium perchlorate

Magnesium permanganate

Magnesium peroxide

Manganese dioxide

Manganese nitrate

Manganous nitrate
Mercuric nitrate (Mercury nitrate)
Mercurous chlorate (Light-sensitive)
Mercurous chromate (Mercury chromate)
Mercurous nitrate
Mercury chlorate
Mercury chromate
12-Molybdosilic acid
Nickel nitrate (Nickelous nitrate)
Nickel nitrite
Niter
Niterox*
Nitrates, inorganic, n.o.s.
Nitric acid (Concentrated, Fuming)
Nitrites, inorganic, n.o.s.
Nitro carbon nitrate
Nitrogen dioxide
Nitrogen peroxide
Nitrogen tetroxide
Nitrogen trichloride
Nitrogen trifluoride
Nitrogen trioxide
Nitronium perchlorate
Nitrosyl chloride
Nitrosyl fluoride
Nitryl chloride
Nitryl fluoride
Osmium amine nitrate
Osmium amine perchlorate
Oxidizing agents/Oxidizers, strong, n.o.s.

Oxone*

Oxygen

Oxygen-17,18

Oxygen difluoride (Oxygen fluoride)

Ozone

Palladium nitrate

Perbromic acid

Percarbonates, inorganic, n.o.s.

Perchlorate, n.o.s.

Perchlorates, inorganic, n.o.s.

Perchloric acid

Perchloryl fluoride

Performic acid

Periodic acid

Permanganate, n.o.s.

Permanganates, inorganic, n.o.s.

Peroxides, inorganic, n.o.s.

Peroxysulfuric acid

Persulfates (Persulphates), inorganic, n.o.s.

Phosphorus oxybromide (Phosphoryl bromide)

Phosphorus oxychloride (Phosphoryl chloride)

Phosphotungstic acid

Pittclor*

Potassium bromate

Potassium chlorate

Potassium chlorite

Potassium chlorochromate

Potassium dichloroisocyanurate

Potassium dichloro-s-triazinetrione, dry

Potassium dichromate

Potassium hypochlorite

Potassium hypochlorite in basic solution

Potassium manganate

Potassium nitrate

Potassium nitrite

Potassium percarbonate

Potassium perchlorate

Potassium periodate

Potassium permanganate

Potassium peroxide

Potassium persulfate (Potassium persulphate)

Potassium superoxide

2-Propanol nitrate

Propyl nitrate

Red lead

Ruthenium tetroxide

Saltpeter

Selenium dioxide

Silver acetate (Fairly weak)

Silver chlorate

Silver chromate

Silver dichromate

Silver nitrate

Silver oxide

Silver perchlorate

Silver permanganate

Silver peroxide

Sodium bichromate

Sodium bromate

Sodium carbonate peroxide

Sodium chlorate

Sodium chlorite

Sodium dichloroisocyanurate

Sodium dichloro-*s*-triazinetrione

Sodium dichromate

Sodium dioxide

Sodium hypochlorite

Sodium hypochlorite in basic solution

Sodium iodate

Sodium 12-molybdosilicate in water

Sodium nitrate (Soda niter, Caliche)

Sodium nitrite

Sodium perborate (Anhydrous and hydrate)

Sodium percarbonates

Sodium perchlorate

Sodium periodate

Sodium permanganate

Sodium peroxide

Sodium peroxoborate

Sodium persulfate (Sodium persulphate)

Sodium pyrophosphate peroxide

Sodium superoxide

Sodium 12-tungstophosphate

Strontium bromate

Strontium chlorate

Strontium nitrate

Strontium perchlorate

Strontium peroxide

Strontium-potassium chlorate

Sulfur dichloride

Sulfur dioxide

Sulfuric acid (Oleum; Oxidizes organics)

Sulfuric acid, concentrated (Mists)

Sulfuric anhydride

Sulfurous acid (Oxidizes limited materials)

Sulfur trioxide (Sulfuric anhydride, Sulphur trioxide)

Sulfurous acid

Sulfur trioxide

Terbium nitrate

Tetrafluorohydrazine

Tetranitromethane (Powerful)

Textone*

Thallium chlorate

Thallium nitrate

Thorium nitrate

Trichloroisocyanuric acid

Trichloro-s-triazinetrione, dry

(mono)-Trichloro-tetra(monopotassium dichloro)penta-s-triazinetri-
one, dry

Uranium nitrate (Uranyl nitrate)

Uranyl nitrate hexahydrate

Urea nitrate

Xenic acid

Xenon compounds (Oxides, Oxyfluorides)

Xenon trioxide, dry

Yellow salt

Zinc ammonium nitrite

Zinc bromate

Zinc chlorate

Zinc dioxide

Zinc nitrate

Zinc perborate

Zinc permanganate

Zinc peroxide

Zirconium nitrate

Zirconyl nitrate (Basic)

RGN 65 REDUCING AGENTS, STRONG OR SPECIAL TYPES OF REDUCING AGENTS

A number of these compounds form toxic and flammable fumes in acid and/or during heating. Some compounds are flammable, while others ignite spontaneously (pyrophoric) in air. A number of the reducing agents evolve hydrogen on contact with water and tend to be explosive. Many compounds react vigorously (explosively) and/or ignite spontaneously on exposure to oxidizing agents. Do not mix many of these compounds with oxidizing agents (RGN 64 and RGN 30). A limited number of reducing agents react under special conditions and/or with specific chemicals. Do not breathe the vapors of these chemicals or the products of reaction and/or decomposition. Many of the metal sufides are reducing agents and can be found in RGN 33. Most nonsoluble sulfides are not listed in this group (RGN 65). Experienced individuals should handle these chemicals properly. **Note**: *Not every reducing agent is identified in this list.*

Aluminum borohydride

Aluminum calcium hydride

Aluminum carbide

Aluminum diethyl monochloride

Aluminum ethylate (Aluminum ethoxide)

Aluminum hydride

Aluminum hypophosphite

Ammonium bisulfide

Ammonium hydrosulfide (Ammonium hydrosulphide)

Ammonium hypophosphite

Ammonium phosphite

Ammonium polysulfide (Ammonium polysulphide)

Ammonium sulfide (Ammonium sulphide)

Ammonium sulfite

Antimony pentasulfide

Antimony sulfide (Antimony sulphide)

Antimony trisulfide (Antimony trisulphide)

Arsenic disulfide

Arsenic pentasulfide

Arsenic sulfide (Arsenic sulphide)

Arsenic trisulfide (Arsenic trisulphide)

Arsine

Barium carbide

Barium hydride

Benzyl silane

Benzyl sodium

Beryllium hydride

Beryllium sulfide

Beryllium tetahydroborate

Bismuth sulfide

Bismuth trisulfide (Bismuth trisulphide)

Boron arsenotribromide

Boron trisulfide

Bromodiborane

Bromosilane

Butyl dichloroborane

n-Butyllithium

Cadmium acetylide

Cadmium sulfide

Calcium carbide

Calcium hydride

Calcium hypophosphite

Calcium sulfide

Carbon (When purifying metals)

Cerium-iron alloys

Cerium

Cerium hydride

Cerium trisulfide

Cerous phosphide

Cesium carbide

Cesium hexahydroaluminate

Cesium hydride

Cesium sulfide

Chlorodiborane

Chlorodiisobutyl aluminum

Chlorodimethylamine diborane

Chlorodipropyl borane

Chlorosilane(s)

Chromium sulfide (Chromic sulfide)

Chromous chloride

Chromous oxalate

Copper acetylide (Cuprous acetylide)

Diamine

DIBAL-H

Diborane

Dicopper(I) acetylide

Diethyl aluminum hydride
Dihydrazine sulfate
Diisobutylaluminum hydride (DIBAL-H)
1,1-Dimethylhydrazine (UDMH)
Dimethyl magnesium
Ethyllithium
Ferric hypophosphite
Formic acid
Geranium sulfide
Gold acetylide
Gold sulfide
Hexaborane
Hexahydride diborane
Hydrazine (Anhydrous)
Hydrazine hydrate
Hydrazine, n.o.s.
Hydrazine sulfate
Hydrides, metal, n.o.s.
Hydrogen, gas
Hydrogen phosphide
Hydrogen selenide
Hydrogen sulfide
Hydrogen telluride
Hydroxylamine
Hydroxylamine hydrochloride
Hydroxylamine acid sulfate (HAS)
Hydroxylamine sulfate (Hydroxylamine sulphate, HS)
Hypophorous acid
Lithium aluminum deuteride (LAD)
Lithium aluminum hydride (LAH)

Lithium aluminum hydride, etheral

Lithium borohydride

Lithium deuteride

Lithium hydride

Lithium nitride

Lithium sulfide

Magnesium (Fine powder, Flakes, Pellets, Turnings, Ribbon, etc.)

Magnesium sulfide

Manganese hypophosphite

Manganese sulfide

Methyl hydrazine (MMH)

Methylphenyldichlorosilane

Nickel subsulfide

Organoborane, n.o.s.

Pentaborane

Phenylhydrazine hydrochloride

Phosphine

Phosphonium iodide

Phosphoric sulfide

Phosphorus (Red amorphous)

Phosphorus (White or yellow)

Phosphorus heptasulfide

Phosphorus pentasulfide

Phosphorus sesquisulfide

Phosphorus trisulfide

Potassium arsenite

Potassium borohydride

Potassium dithionite (Potassium hydrosulfite)

Potassium hydride

Potassium hydrosulfite

Potassium hypophosphite

Potassium sulfide, anhydrous

Potassium sulfide, with water of hydration or crystallization

Red arsenic

Red phosphorus

Reducing agents, strong, n.o.s.

Silver acetylide

Silver sulfide

Sodium aluminum hydride

Sodium bis(2-methoxyethoxy)aluminohydride

Sodium dithionite

Sodium hydride

Sodium hydrosulfite (Sodium hydrosulphite)

Sodium hypophosphite

Sodium hyposulfite

Sodium-lead alloy

Sodium sulfide (Sodium sulphide)

Sodium tetrasulfide

Stannic sulfide

Stannous acetate

Strontium monosulfide

Strontium tetrasulfide

Sulfur dioxide

Tetraborane

Tetraphosphorus trisulfide

Thallium sulfide

Titanium hydride

Titanium sesquisulfide

Titanium sulfide

Titanium trichloride (Pyrophoric)

Tri-*n*-butyl borane

Tributyl phosphine (Tributyl phosphane)

Triethyl antimony

Triethylstibine

Trimethyl antimony

Trimethylstibine

Uranium hydride

Uranium sulfide

Vanadium dichloride

Vanadous chloride

White phosphorus

Yellow phosphorus

Zepar* BP

Zinc acetylide

Zinc dithionite (Zinc hydrosulfite)

Zirconium hydride

RGN 66 AIR-REACTIVE, AIR-SENSITIVE, AND SELF-REACTIVE SUBSTANCES

Several compounds are temperature-sensitive and self-decompose. Some compounds ignite spontaneously in air and are known as "pyrophoric." Many compounds are self-reactive or highly flammable in air. Self-ignition and/or self-decomposition may be triggered by heat, chemical reaction, friction, impact, contamination, air, oxygen, sparks, flame, grinding, rubbing, etc. Decomposition may be self-accelerating and produce large volumes of gases, thus providing the opportunity for an explosion. Decomposition and/or reaction of these substances may produce irritating, toxic, and/or corrosive gases which may cause injury or death. These compounds often re-ignite after a fire is extinguished. Some compounds will explode. Also, powders, dusts, or vapors can explode in air. Each compound must be stored at a specific temperature or under controlled conditions to avoid decomposition, self-ignition, or reaction. Some compounds are both air sensitive (oxidize or react in air) and water sensitive. Many compounds are stored under nitrogen, inert gas, or other nonreactive materials. Special transfer and handling conditions are required with most of these chemicals. Experienced individuals should handle these chemicals properly. **Note**: *Not every substance or chemical that is air reactive, air-sensitive, or self-reactive is included in this list.*

Acetone cyanohydrin

Alkali metal alcoholates (Self-heating)

Aluminum alkyl (Aluminum trialkyl), n.o.s.

Aluminum borohydride

Aluminum hypophosphite

Ammonium hypophosphite

Amyl nitrite (Isoamyl nitrite)

Arsenic trifluoride (Arsenic fluoride)

Azodicarbonamide (1,1'-Azobisformamide)

2,2'-Azodi-(2,4-dimethyl-4-methoxyvaleronitrile)

2,2'-Azodi-(2,4-dimethylvaleronitrile)

1,1'-Azodi-(hexahydrobenzonitrile)

Azodiisobutyronitrile

2,2'-Azodi-(2-methylbutronitrile)

Barium iodide

Benzene-1,3-disulfohydrazide (Benzene-1,3-disulphohydrazide)

Benzenephosphorus dichloride

Benzene sulfohydrazide (Benzene sulphohydrazide)

4-[Benzyl(ethyl)amino]-3-ethoxybenzenediazonium zinc chloride

4-[Benzyl(methyl)amino]-3-ethoxybenzenediazonium zinc chloride

Bismuth ethyl chloride

Borane

Boron, dust

Bromine chloride

N-Bromosuccinimide, dry (NBS, Use respirator)

1,3-Butadiene

Butylene oxide

n-Butyllithium

Butylmagnesium chloride

Butyltin chloride

5-tert-Butyl-2,4,6-trinitro-m-xylene

Calcium hydrosulfide

Calcium resinate (Spontaneous heating)

Calcium silicide

Calcium silicon (Calcium-silicon alloy)

Charcoal wood

3-Chloro-4-diethylaminobenzenediazonium zinc chloride

Chloroformoxime

Cobalt

Cobalt resinate (Cobaltous resinate)

Cotton, acetylated

Cyanogen azide

Decaborane

2-Diazo-1-naphthol-4-sulfochloride (2-Diazo-1-naphthol-4-sulphochloride)

2-Diazo-1-naphthol-5-sulfochloride (2-Diazo-1-naphthol-5-sulphochloride)

DIBAL-H

Dichloroformoxime

2,5-Diethoxy-4-morpholinobenzenediazonium zinc chloride

Diethylaluminum chloride (DEAC)

Diethylaluminum hydride

Diethylcadmium

p-Diethylnitrosoaniline

Diethylzinc (Pyrophoric)

Diiododiethyl sulfide

Diisobutylaluminum hydride

Diisopropyl peroxydicarbonate (Isopropyl percarbonate)

Dimethoxyphenyllithium

4-Dimethylamino-6-(2-dimethylaminoethoxy)toluene-2-diazonium zinc chloride

N,N'-Dinitroso-N,N'dimethylterephthalamide

N,N'-Dinitrosopentamethylenetetramine (DNPT)

Diodacetylene

Diphenyloxide-4,4'-disulfohydrazide (Diphenyloxide-4,4'-disulpho-hydrazide)

4-Dipropylaminobenzenediazonium zinc chloride

Ethyl aluminum dichloride (EADC)

Ethyl aluminum sesquichloride (EASC)

Ethylhexaldehyde

Ethyl iodoacetate (Light and air sensitive)

Ethyllithium

Ethyltrichlorosilane

Europium, powder

Glyoxal (Vapor and air mixture *may* explode)

Grignard reagents (May react with air)

3-(2-Hydroxyethoxy)-4-pyrrolidin-1-yl benzenediazonium zinc chloride

Hydroxylamine

Indene

Iodized oil (Oxidizes in air and light)

Iron pentacarbonyl (Iron carbonyl)

Isopropyl percarbonate

Lithium nitride

Lithium stearate (Dust; spontaneously combustible)

Magnesium amide

Maneb (Manex*)

Methylaluminum sesquibromide

Methylaluminum sesquichloride

Methyl hydrazine (May self ignite in air)

Methyllithium

Methyltrichlorosilane

Molybdenum pentachloride

Musk xylene (Self-reactive perfume)

Nickelocene

Octyl iodide (Light and air sensitive)

Oxygen difluoride (Oxygen fluoride), gas

Pentaborane

Phenylmercaptan

Phenylphosphine

Phosphorus, red

Phosphorus, white or yellow (Dry; Store under water or in solution)

Phosphorus acid (Absorbs oxygen readily)

Phosphorus pentafluoride

Phosphorus pentasulfide

Pyrethrins I

Radium-226 element

Red phosphorus

Salicylic acid dust (Explosive mixtures in air)

Silane (Silicon tetrahydride)

Silyl compounds, n.o.s.

Sodium 2-diazo-1-naphthol-4-sulfonate (Sodium 2-diazo-1-naphthol-4-sulphonate)

Sodium 2-diazo-1-naphthol-5-sulfonate (Sodium 2-diazo-1-naphthol-5-sulphonate)

Sodium thioglycolate

Stannic bromide (Fumes in air)

Strontium iodide (Turns yellow in air or light, Decomposes in moist air)

Tetrasilane

Thallium monoxide

Thallium oxide

Thioxylenol

TIBAL

Titanium dichloride

Titanium trichloride

Tri-*n*-butylaluminum

Tri-*n*-butylborane (Pyrophoric)

Trichloromethyl chloroformate (Poison; Easily decomposed)

Trichloronitrosomethane (Poison, slowly decomposes)

Triethylaluminum (ATE, TEA)

Triethylborane

Tri-*n*-hexylaluminum

Triisobutylaluminum (TIBAL)

Trimethylaluminum

Tri-*n*-octylaluminum

Tri-*n*-propylaluminum

Trisilane (Explodes on contact with air)

Uranium hydride

Uranium monocarbide

Vanadium tetrachloride

White phosphorus

Xanthic acid

Yellow phosphorus

Zinc ethyl (Diethylzinc; Pyrophoric)

Zirconium carbide (Dust or powder)

RGN 67 WATER-REACTIVE AND MOISTURE-SENSITIVE COMPOUNDS

The organic chemicals are flammable and tend to decompose and/or ignite in water. Several water-reactive organic compounds form organic acids with exposure to moisture and water.

Whereas, some metals, amalgams, and/or metal compounds will ignite when exposed to water. Several metal halides and metal compounds tend to decompose in water. Many chemicals tend to explode or react vigorously with water to cause damage to the surroundings and humans. A number of these chemicals exposed to water (or moisture in air) form flammable gases and/or toxic gases.

Several chemicals in this group should not be mixed with strong reducing agents (RGN 65). Do not breath the vapors of these chemicals and avoid the reaction products of these chemicals. Experienced individuals should handle these chemicals properly. **Note:** *Not every substance that is water-reactive or moisture-sensitive is included in this list. RGN 21 includes metals and alloys that are water-reactive.*

Acetic anhydride

Acetyl bromide

Acetyl chloride

Acetyl iodide

Acetyl ketene (Diketene; Uninhibited/partially inhibited)

Alkali metal(s), n.o.s.

Alkali metal amalgam, n.o.s. (Liquid or Solid)

Alkali metal amide

Alkaline-earth metal alcoholate, n.o.s.

Alkaline-earth metal alloy (Amalgams)

Alkyl aluminum chloride

Allyl trichlorosilane

Aluminum alkyl (Aluminum trialkyl), n.o.s.

Aluminum alkyl halide, n.o.s.

Aluminum alkyl hydride, n.o.s.

Aluminum aminoborohydride

Aluminum borohydride

Aluminum bromide

Aluminum calcium hydride

Aluminum carbide

Aluminum chloride, anhydrous

Aluminum diethyl monochloride

Aluminum dross

Aluminum ferrosilicon powder

Aluminum hydride

Aluminum iodide, anhydrous

Aluminum nitride

Aluminum phosphide

Aluminum phosphide pesticide

Aluminum sulfide

Aluminum tetrahydroborate

Aminosalicylic acid (PASA, PAS)

Ammonium picrate

Amyltrichlorosilane

Anisoyl chloride

Antimony chloride

Antimony fluoride

Antimony pentachloride
Antimony pentafluoride
Antimony sulfate
Antimony tribromide
Antimony trichloride
Antimony triiodide
Antimony trivinyI
Aqua regia
Arsenic bromide (Arsenic tribromide)
Arsenic chloride (Arsenic trichloride)
Arsenic fluoride (Arsenic trifluoride; Fumes in air)
Arsenic iodide (Arsenic triiodide)
Arsenic pentafluoride, gas
Arsenic tribromide
Arsenic trichloride
Arsenic trifluoride
Arsenic triiodide
Barium carbide
Barium dichromate
Barium oxide (Barium monoxide)
Barium selenide
Barium silicide
Barium sulfide (Barium sulphide)
Benzenephosphorus dichloride
Benzenephosphorus oxydichloride
Benzenephosphorus thiodichloride
Benzenesulfonyl chloride (Benzenesulphonyl chloride)
Benzotrichloride
Benzoyl chloride
Benzyl chlorocarbonate

Benzyl chloroformate

Benzyl silane

Benzyl sodium

Beryllium hydride

Beryllium tetahydroborate

Bismuth bromide

Bismuth chloride

Bismuth pentafluoride

Bis(trichlorosilylethane)

Borane

Boron phosphide

Boron tribromide

Boron trichloride

Boron trifluoride acetic acid complex

Boron trifluoride dimethyl etherate

Boron trifluoride propionic acid complex

Bromine chloride

Bromine monofluoride

Bromine pentafluoride

Bromine trifluoride

Bromoacetic acid

Bromodiethylaluminum

Bromophosgene

Butyldichloroarsine

Butyl isocyanate (*n-*, *iso-*, *tert-*)

n-Butyllithium (In moist air)

Butylmagnesium chloride (In ether)

Butyltrichlorosilane

Butyric anhydride

Cadmium acetylide (In slightly acidified water)

Cadmium amide

Calcium carbide

Calcium cyanamide

Calcium dithionite

Calcium hydride

Calcium manganese silicon

Calcium metal (Crystalline)

Calcium nitride

Calcium phosphide

Calcium silicide (Decomposes in hot water)

Calcium silicofluoride (Decomposes in hot water)

Calcium sulfide (Partial decomposition)

Carbon suboxide

Carbonyl cyanide

Carbonyl fluoride

Cesium amide

Cesium hydride

Cesium phosphide

Cesium tetroxide

Cesium trioxide

Chloracetyl chloride

Chloramine (Not to be confused with Chloramine-T)

Chlorine dioxide

Chlorine dioxide, hydrate (Frozen)

Chlorine heptoxide

Chlorine monofluoride

Chlorine pentafluoride

Chlorine trifluoride

Chloroacetyl chloride

o-Chlorobenzotrichloride

Chlorochromic anhydride
Chlorodiisobutyl aluminum
Chloromethylchloroformate
Chloropentafluoroacetone
Chlorophenyltrichlorosilane
Chloropivaloyl chloride
Chlorosilane
Chlorosulfonic acid
Chlorotrifluoroethylene
Chlorovinyldichloroarsine
Chlorovinylmethylchloroarsine
Chromic anhydride
Chromium oxychloride
Chromyl chloride
Cinnamoyl chloride
Cobalt fluoride (Cobaltic trifluoride)
Copper acetylide (Cuprous acetylide; Slightly acidified solution)
Copper phosphide
Cyanogas*
Cyanuric chloride
Cyclohexenyltrichlorosilane
Cyclohexyl trichlorosilane
Cyclooctadiene phosphines
Decaborane
Diacetylene, gas or liquid (Moist silver salts)
Dibenzyldichlorosilane
Diborane
Dibromomethyl ether
Dichloroacetyl chloride
Dichlorodimethylsilane

Dichloroethylarsine
Dichloromethylchloroformate
Dichlorophenyl isocyanate
Dichlorophenyltrichlorosilane
Dichlorosilane
Dicopper(I) acetylide
Diethylaluminum chloride (DEAC)
Diethylaluminum hydride
Diethyldichlorosilane
Diethylgermanium dichloride
Diethylthiophosphoryl chloride
Diethylzinc
Diisopropyl beryllium
Diisopropyl fluorophosphate (DFP)
Dimethylcarbamoyl chloride
Dimethyl chlorothiophosphate
Dimethyldichlorosilane
Dimethyldiethoxysilane
Dimethyl magnesium
Dimethylthiophosphoryl chloride
Dimethylzinc
Diphenylchloroarsine
Diphenyldichlorosilane
Diphenylmethane-4,4'-diisocyanate (MDI)
Disulfuryl chloride (Pyrosulfuryl chloride)
Dodecyltrichlorosilane
Ethanoyl chloride (Acetyl chloride)
Ethoxycarbonyl isothiocyanate
Ethylaluminum dichloride (EADC)
Ethyl aluminum sesquichloride (EASC)

Ethyl borate

Ethyl chloroformate (Ethyl chlorocarbonate)

Ethyl chlorosulfonate

Ethyl chlorothioformate

Ethyl dichloroarsine

Ethyl dichlorosilane

Ethyl isocyanate

Ethyllithium

Ethyl nitrite

Ethyl oxalate (Slow decomposition)

Ethylphenyldichlorosilane

Ethyl phosphonous dichloride, anhydrous

Ethylsulfuric acid (Ethylsulphuric acid)

Ethyltrichlorosilane

Europium, powder

Ferrosilicon

Ferrous phosphide

Fluorine monoxide

Fluorosulfonic acid (Fluorosulphonic acid)

Fluosulfonic acid (Fluosulphonic acid, Fluorosulfuric acid))

Formyl fluoride

Fumaryl chloride

Furoyl chloride (Lachrymator)

Germanium dichloride

Germanium tetrachloride

Gold acetylide

Grignard reagent(s)

Hexadecyltrichlorosilane

Hexafluoroacetone

Hexahydride diborane (Diborane Hexahydride)

Hexamethylene diisocyanate

Hexanoyl chloride

Hexyltrichlorosilane

Hydrazine perchlorate

Hydrides, metal, n.o.s.

Hydriodic acid (Hydrogen iodide; Concentrated solution)

Hydrobromic acid (Hydrogen bromide; Concentrated solution)

Hydrochloric acid (Hydrogen chloride; Concentrated solution)

Hydrogen bromide, gas, anhydrous

Hydrogen chloride, gas, anhydrous

Hydrogen iodide, gas, anhydrous

Hydrogen telluride

Iodine monobromide (Iodine bromide)

Iodine monochloride (Iodine chloride)

Iodine pentafluoride

Iodine trichloride

Iodoacetic acid, sodium salt

N-Iodosuccinimide

IPDI

Isoamyldichloroarsine

Isobutyl chloroformate

Isobutyl isocyanate

Isobutyric anhydride

Isocyanate solution, n.o.s.

Isocyanate(s), n.o.s.

Isocyanatobenzotrifluoride(s)

Isophorone diisocyanate (IPDI)

Isophthaloyl chloride

Isopropyl chloroacetate

Isopropyl chloroformate

Isopropyl isocyanate

K-Selectride* (Tri-*sec*-butylborohydride in tetrahydrofuran)

LAD

LAH

LATB

Lead selenide

Lewisite

Lime, chlorinated (Evolves chlorine)

Lithium acetylide-Ethylenediamine complex

Lithium alkyls

Lithium aluminum deuteride (LAD)

Lithium aluminum hydride (LAH)

Lithium aluminum hydride, etheral

Lithium aluminum tri-*tert*-butoxyhydride (LATB)

Lithium amide

Lithium borohydride

Lithium carbide

Lithium deuteride

Lithium ferrosilicon

Lithium hydride (In moist air)

Lithium nitride

Lithium silicon

L-Selectride*

Magnesium alkyls

Magnesium aluminum phosphide

Magnesium amide

Magnesium diamide

Magnesium diphenyl

Magnesium hydride

Magnesium phosphide

Magnesium silicide

Maleic acid

Maleic anhydride

Metal alkyl halides, n.o.s.

Metal alkyls, n.o.s.

Metal aryl halides, n.o.s.

Metal aryl hydrides, n.o.s.

Metal aryls, n.o.s.

Metal hydrides, n.o.s

Methanesulfonyl chloride (Methanesulphonyl chloride)

Methoxymethyl isocyanate

Methylaluminum sesquibromide

Methylaluminum sesquichloride

Methyl bromoacetate

Methyl chloroacetate

Methyl chloroformate

Methyl chloromethyl ether

Methylchlorosulfonate

Methyl cyanoformate

Methyldichloroacetate

Methyldichloroarsine

Methyl dichlorosilane

Methylene diisocyanate

Methyl fluorosulfonate

Methyl isocyanate

Methyl lactate

Methylmagnesium bromide

Methylmagnesium chloride

Methylmagnesium iodide

Methyl orthosilicate

Methylphenyldichlorosilane

Methyl phosphonic dichloride

Methyl phosphonous dichloride

Methyl trichloroacetate

Methyltrichlorosilane

Methylvinyldichlorosilane

Molybdenum pentachloride

Muriatic acid (concentrated)

Nickel antimonide

Nitrating acid

Nitric acid (Concentrated, Fuming)

Nitrobenzoyl chloride

Nitrogen trichloride

Nitrohydrochloric acid (Aqua regia)

p-Nitrosodiethylaniline

p-Nitrosodimethylaniline

Nitrosyl chloride

Nitrosylsulfuric acid (Nitrosylsulphuric acid)

Nitryl chloride

Nonyltrichlorosilane

Octadecyltrichlorosilane

Octanoyl chloride

Octyl trichlorosilane

Oleum (Oil of vitriol, Concentrated sulfuric acid)

Organoborane, n.o.s.

Organometallic compound, n.o.s (water-reactive)

Oxalyl chloride (Ethanedioyl chloride)

Oxygen fluoride (Oxygen difluoride)

Pelargonyl chloride

Pentaborane

Perchloromethyl mercaptan
Phenylacetyl chloride
Phenyl chloroformate
Phenyl isocyanate
Phenylphosphorus dichloride
Phenylphosphorus thiodichloride
Phenyltrichlorosilane
9-Phosphabicylononane(s)
Phosphonium iodide
Phosphoric anhydride
Phosphoric bromide
Phosphoric chloride
Phosphoric sulfide
Phosphorous heptasulfide
Phosphorous acid (Phosphonic acid, ortho)
Phosphorous oxybromide
Phosphorous oxychloride
Phosphorus pentabromide (Phosphoric bromide)
Phosphorus pentachloride (Phosphoric chloride)
Phosphorus pentafluoride
Phosphorus pentasulfide
Phosphorus pentoxide
Phosphorus sesquisulfide
Phosphorus tribromide
Phosphorus trichloride
Phosphorus triiodide
Phosphorus trisulfide
Phosphoryl bromide
Phosphoryl chloride
Phthalic anhydride

Phthaloyl chloride

Polyphenyl polymethylisocyanate

Potassium amide

Potassium dithionite

Potassium hydride

Potassium hydrosulfite

Potassium oxide

Potassium percarbonate (Light and moisture sensitive)

Potassium peroxide

Potassium phosphide

Potassium sulfide, anhydrous (May ignite spontaneously; Dust or powder is explosive)

Potassium sulfide, with small amount of water of hydration or crystallization (May ignite)

Propionic anhydride

Propionyl chloride

Propyl chloroformate

Propyl isocyanate

Propyltrichlorosilane

Pyromellitic dianhydride (PMDA)

Pyrosulfuryl chloride (Pyrosulphuryl chloride)

Radium-226 element

Selenium oxychloride

Selenium tetrafluoride

Silane (Silicon tetrahydride)

Silicochloroform

Silicon tetrabromide

Silicon tetrachloride

Silicon tetrafluoride

Silicon tetrahydride

Silver acetylide

Silyl compounds, n.o.s.

Sisal dust

Sodium aluminum hydride

Sodium amide (Sodamide)

Sodium bis(2-methoxyethoxy)aluminohydride

Sodium borohydride

Sodium cuprocyanide (Sodium copper cyanide)

Sodium dithionite (Sodium hydrosulfite)

Sodium ethylate (Sodium ethoxide)

Sodium hydride

Sodium hydrosulfite (Sodium hydrosulphite)

Sodium methoxide

Sodium methylate

Sodium monoxide (Sodium oxide)

Sodium peroxide

Sodium persulfate (Sodium persulphate)

Sodium phosphide

Stannic chloride, anhydrous

Stannic phosphide

Strontium phosphide

Sulfonyl chloride

Sulfonyl fluoride

Sulfur bromide

Sulfur chloride

Sulfur dichloride

Sulfuric acid (Oleum; Concentrated)

Sulfuric acid, concentrated (Mists)

Sulfuric anhydride

Sulfur monochloride

Sulfur oxychloride

Sulfur pentafluoride

Sulfur tetrafluoride

Sulfur trioxide (Sulfuric anhydride)

Sulfuryl chloride

Sulfuryl fluoride

Sulphur chloride(s)

Sulphuric acid

Sulphur tetrafluoride

Sulphur trioxide (Sulfuric anhydride)

Sulphuryl chloride

Tellurium dibromide

Tellurium dichloride

Tellurium tetrabromide

Terephthaloyl chloride

Tetrabutyl titanate (TBE)

Tetracyanoethylene

Tetrahydrophthalic anhydride(s)

Tetramethylammonium chlorodibromide

Tetraphosphorus trisulfide

Thallium monoxide

Thiocarbonyl chloride

Thionyl chloride

Thiophosgene

Thiophosphoryl chloride

Thiourea dioxide

Thorium carbide

Thorium fluoride

TIBAL

Tin tetrachloride (Tin chloride, Stannic chloride)

Titanium dichloride

Titanium disulfide (Titanium disulphide)

Titanium tetrachloride

Titanium trichloride (Pyrophoric)

Titanocene dichloride

Toluene-2,4-diisocyanate (TDI)

Toluene-2,6-diisocyanate

Triallylborate

Tri-n-butylaluminum (Pyrophoric)

Tributyl phosphite

Trichloroacetyl chloride

Trichloroborane

Trichlorosilane

Triethylaluminum (ATE, TEA)

Triethyl antimony

Triethyl arsine

Triethylborane

Triethylborate

Triethylorthoformate

Triethyl stibine

Triisobutylaluminum (TIBAL)

Trimethylaluminum (ATM)

Trimethyl antimony

Trimethyl arsine

Trimethyl borate

Trimethylchlorosilane

Trimethylhexamethylene diisocyanate

Trimethylstibine

Trioctylaluminum

Tripropyl stibine

Trisilyl arsine
Trivinyl stibine
Tungsten hexachloride
Tungsten hexafluoride, gas or liquid
Tungsten oxychloride
Uranium hexafluoride (Radioactive risk)
Vandium oxytrichloride
Vandium tetrachloride
Vandium trichloride
Vikane*
Vinyl chloroacetate
Vinyl trichlorosilane
Water reactive/Water sensitive substances, n.o.s.
Zinc acetylide
Zinc antimonide
Zinc dioxide
Zinc dross
Zinc ethyl (Diethyl zinc; Pyrophoric)
Zinc perborate
Zinc peroxide
Zinc phosphide
Zinc propionate
Zinc residue and skimmings
Zinc selenide (Fire risk)
Zinc telluride
Zirconium hydride
Zirconium tetrachloride
Zirconocene dichloride

INDEX

ABOUT THE AUTHOR

Donald A. Drum was a Professor of Forensic Science, Organic Chemistry, Environmental Science, Chemistry, and Physics. Professor Drum was a member of the American Chemical Society, the Air and Waste Management Association, and the American Society for Engineering Education. In addition, he has taught professional seminars to industry employees, investigators and security agents, government officials, and members of the transportation sector. He is the co-editor of McGraw-Hill's *Environmental Field Test Ready Reference Handbook*.

www.ingramcontent.com/pod-product-compliance
Lightning Source LLC
Chambersburg PA
CBHW011225210326
41598CB00039B/7311